Springer Series in Operations Research

Editors:
Peter W. Glynn   Stephen M. Robinson

Springer

New York
Berlin
Heidelberg
Hong Kong
London
Milan
Paris
Tokyo

Springer Series in Operations Research

Kathrin Klamroth

Single-Facility Location Problems with Barriers

With 61 Illustrations

 Springer

Kathrin Klamroth
Institute of Applied Mathematics
University of Erlangen-Nuremberg
Martensstrasse 3
91058 Erlangen
Germany
klamroth@am.uni-erlangen.de

Series Editors:

Peter W. Glynn
Department of Management Science and
Engineering
Terman Engineering Center
Stanford University
Stanford, CA 94305-4026
USA
glynn@leland.stanford.edu

Stephen M. Robinson
Department of Industrial Engineering
University of Wisconsin–Madison
1513 University Avenue
Madison, WI 53706-1572
USA
smrobins@facstaff.wisc.edu

Library of Congress Cataloging-in-Publication Data
Klamroth, Kathrin.
    Single-facility location problems with barriers / Kathrin Klamroth.
        p.   cm. — (Springer series in operations research)
    Includes bibliographical references and index.

    1. Location problems (Programming)   I. Title.   II. Series.
    QA402.6 .K53   2002
    519.7—dc21   2002067017

ISBN 978-1-4419-3027-9   e-ISBN 978-0-387-22707-8

Printed in the United States of America.

9 8 7 6 5 4 3 2 1

www.springer-ny.com

Springer-Verlag   New York Berlin Heidelberg
A member of BertelsmannSpringer Science+Business Media GmbH

Preface

Everyday life bears a multitude of location problems and locational decisions. These may be as simple as how best to place a pencil on a desk without having to reach too far and still keeping the work space clear, up to the question of where to place the next out of hundreds of thousands of transistors on a microchip. Some of these questions have easy answers, while others are so complex that not even satisfactory solutions are available, never mind asking for optimized placement. The scales of problems reach from microchip design up to global trade and may demand consideration of one, two, three, or even more dimensions.

As modern life encounters an ever increasing concentration in many respects, usually a multitude of restrictions will be imposed on a problem. These restrictions may be classified as regions of limited or forbidden placement of a new facility or as regions with limitations on traveling. Areas where the placement of a new facility is forbidden, referred to as *forbidden regions*, can be used to model, for example, protected areas or regions where the geographic characteristics forbid the construction of the desired facility. Limitations on traveling are constituted by *barrier regions* or *obstacles* like military regions, mountain ranges, lakes, big rivers, interstate highways, or, on smaller scales, machinery and conveyor belts in an industrial plant.

The present text develops the mathematical implications of barriers to the geometrical and analytical characteristics of continuous location problems. Besides their relevance in the application of location-theoretic results, *location problems with barriers* are interesting from a mathematical point of view. The nonconvexity of distance measures in the presence of barriers leads to nonconvex optimization problems. Most of the classical methods in continuous location theory rely heavily on the convexity of the objective function and will thus fail in this context. On the other hand, general methods in global optimization capable of treating nonconvex problems ignore the geometric characteristics of the location problems considered. Theoretic as well as algorithmic approaches are utilized to overcome the described difficulties for the solution of location problems with barriers. Depending on the barrier shapes, the underlying distance measure and the type of objective function, different concepts are conceived to handle the nonconvexity of the problem.

The first part of the book provides the theoretical background for the description of location problems with barriers. Theoretical results as well as algorithmic approaches for location problems with barriers are presented. Starting with a discussion of location problems with polyhedral barriers in a very general setting, the emphasis of Part II is on special barrier shapes, including circular barriers as well as line barriers. Part III covers special distance and objective functions of mathematical as well as practical interest. Particular focus is placed on the cases where more than one decision-maker is involved in the locational decision. Extensions and generalizations of the results presented are pointed out. An application of the developed concepts to a real-world problem concludes the book in Part IV.

In particular, different kinds of distance measures are reviewed in Chapter 1. Barriers are formally introduced and discussed with respect to their impact on the distance measure in Chapter 2. Chapter 3 gives a brief overview of planar location models. Location problems with barriers are formally introduced and discussed in the light of the existing literature. In Chapter 4, lower and upper bounds for various types of location problems with barriers are developed. Chapter 5 contains a detailed discussion of location problems with polyhedral barriers. The results and algorithms can be applied to a large variety of problems, including the well-known Weber and center problems. Moreover, an extension to the location of more than one facility is presented at the end of Chapter 5. Chapters 6 and 7 focus on circular barriers and line barriers, respectively, and utilize their properties to develop specialized solution concepts. The impact of the choice of a suitable distance measure on the corresponding location model is outlined in Chapters 8 and 9. In Chapter 8 Weber problems with block norms are discussed. A finite dominating set is constructed allowing for an efficient solution of the problem. Chapter 9 concentrates on center problems with the Manhattan metric and derives a dominating set result for this difficult class of problems. Since many locational decisions, especially in regional and social planning, involve several decision-makers with different priorities, location models including multiple objective functions are discussed in Chapter 10. Finally, the study of the application of the developed methods to a problem in regional planning is presented in Chapter 11.

Although the book is a monograph, parts of it may be used in an advanced course on location analysis. Based on the introductory Part I, the subsequent chapters are mostly self-contained and could individually be selected for an intensive study of a special class of location problems with barriers. Additionally, they could be used as the basis for advanced student projects accompanying or following a location course.

This book would not have been possible without the help of many people contributing in a multitude of ways. First and foremost, I want to thank Horst W. Hamacher. His encouragement and advice were a constant motivation during my work in Kaiserslautern and for the development of the contents of this text. I am also indebted to my colleagues in the group of Mathematical Optimization at the University of Kaiserslautern, and in the Department of Mathematical Sciences at Clemson University, who influenced my work by their ideas and comments. I extend my thanks to the rest of the faculties, staff, and students in Kaiserslautern and in Clemson for all their help and for many fruitful discussions. Particularly I want to thank Matthias Ehrgott and Anita Schöbel for their helpful comments and suggestions on earlier versions of this text. The financial support by the German research foundation DFG (Deutsche Forschungsgemeinschaft) is appreciated. For generous advice, help, and technical support, I thank the Springer-Verlag series editors and staff, namely, Achi Dosanjh, Elizabeth Young, Omar J. Musni, Steve Pisano and David Kramer.

Above all, I thank my parents and Dorel, my husband to be. Without their constant support and encouragement this work would not have been possible.

Erlangen, July 2002   Kathrin Klamroth

Contents

Part I

Introduction and General Results

1
Measuring Distances

The choice of a suitable distance function plays a crucial role for a good estimation of travel distances in realistic environments. Depending on the mode of travel and the type of problem considered, we may use, for example, Euclidean distances to model airtravel or the Manhattan metric for transportation problems in New York. This indicates that different modes of transportation require different ways of distance estimation. Consequently, an increasing effort to estimate travel distances with high accuracy has arisen from the growing variety in the transportation sector.

A well-studied family of distance-predicting functions in urban environments is that of *multiparameter round norms*; see, for example, Love and Morris (1972, 1979, 1988). Applications and comparisons of these norms can be found in Berens (1988), Berens and Koerling (1985, 1988), and Fernández et al. (2002).

An alternative concept was used in Witzgall (1964) and Ward and Wendell (1980, 1985), who suggested *block norms* to model road distances for a continuous model space. Comparisons of both concepts in the context of location modeling are given in Love and Walker (1994), Thisse et al. (1984), Üster and Love (1998a,b), among others. Asymmetric *polyhedral gauges*, which can be viewed as a generalization of block norms, were successfully applied to restricted location problems; see, for example, Nickel (1995). Their general applicability for the approximation of network distances was discussed in Müller (1997).

In the following a brief overview about norms and distance measures used in location modeling is given. For a detailed survey on the estimation of distances the reader may refer to Perreur (1989) and, in the context of location modeling, to Brimberg and Love (1995).

1.1 Norms and Metrics

To prepare a suitable definition of distance functions for the following chapters we use the norm definitions by Minkowski (1911).

Definition 1.1. (Minkowski, 1911) *Let S be a compact and convex set in $\mathbb{R}^n$ containing the origin in its interior. Furthermore, let S be symmetric with respect to the origin and let $X \in \mathbb{R}^n$. The norm $\gamma : \mathbb{R}^n \to \mathbb{R}$ of X with respect to S is then defined as*

$$\gamma(X) := \inf\{\lambda > 0 \, : \, X \in \lambda S\}. \tag{1.1}$$

We will often refer to $\gamma(X)$ as $\|X\|_\gamma$.

Every norm γ defined according to Definition 1.1 satisfies the *norm properties* in $\mathbb{R}^n$:

Lemma 1.1. (Minkowski, 1911) *Let γ be defined according to Definition 1.1. Then for all $X, Y \in \mathbb{R}^n$ and for all $\lambda \in \mathbb{R}$,*

$$\gamma(X) \geq 0 \quad and \quad (\gamma(X) = 0 \Leftrightarrow X = \underline{0}), \tag{1.2}$$

$$\gamma(\lambda X) = |\lambda|\, \gamma(X). \tag{1.3}$$

$$\gamma(X + Y) \leq \gamma(X) + \gamma(Y). \tag{1.4}$$

Given a functional $\gamma : \mathbb{R}^n \to \mathbb{R}$ satisfying (1.2), (1.3), and (1.4), the set S can be written as

$$S = \{X \in \mathbb{R}^n \, : \, \gamma(X) \leq 1\}. \tag{1.5}$$

It is called the *unit ball* or *sublevel set of level 1* with respect to γ (see, for example, Rockafellar, 1970). Since γ satisfies (1.2), (1.3), and (1.4), the unit ball of γ is compact, convex, and symmetric with respect to the origin, and it contains the origin in its interior.

Every norm γ defines a distance measure in $\mathbb{R}^n$ in the following way (see, for example, Forster, 1977):

Definition 1.2. *Let γ be a norm in $\mathbb{R}^n$ and let $X, Y \in \mathbb{R}^n$. Then the metric induced by $\| \bullet \|_\gamma$ for two points X and Y in $\mathbb{R}^n$ is defined by*

$$\gamma(X, Y) := \gamma(Y - X) = \|Y - X\|_\gamma. \tag{1.6}$$

We will often write $d(X, Y)$ instead of $\gamma(X, Y)$ and refer to the norm with the induced metric d as $\| \bullet \|_d$.

Lemma 1.2. *Let d be a metric induced by a norm $\| \bullet \|_d$ in $\mathbb{R}^n$ according to Definition 1.2. Then d satisfies, for all $X, Y, Z \in \mathbb{R}^n$,*

$$d(X,Y) \geq 0 \quad and \quad (d(X,Y) = 0 \Leftrightarrow X = Y), \tag{1.7}$$

$$d(X,Y) = d(Y,X), \tag{1.8}$$

$$d(X,Z) \leq d(X,Y) + d(Y,Z). \tag{1.9}$$

A well-studied family of norms (and of corresponding metrics) is the family of l_p norms, $p \in [1, \infty]$. The following definition is taken from Forster (1983) and can be found in almost every textbook on fundamental analysis:

Definition 1.3. *Let $X \in \mathbb{R}^n$. The l_p norm of $X = (x_1, \ldots, x_n)^T \in \mathbb{R}^n$ is defined as*

$$\|X\|_{l_p} = l_p(X) = \left(\sum_{i=1}^{n} |x_i|^p \right)^{\frac{1}{p}} \quad for\ p \in [1, \infty), \tag{1.10}$$

$$\|X\|_{l_\infty} = l_\infty(X) = \max_{i=1,\ldots,n} |x_i|. \tag{1.11}$$

The corresponding l_p metric in $\mathbb{R}^n$ for two points $X = (x_1, \ldots, x_n)^T$ and $Y = (y_1, \ldots, y_n)^T$ is given by

$$l_p(X,Y) = \left(\sum_{i=1}^{n} |y_i - x_i|^p \right)^{\frac{1}{p}} \quad for\ p \in [1, \infty), \tag{1.12}$$

$$l_\infty(X,Y) = \max_{i=1,\ldots,n} |y_i - x_i|. \tag{1.13}$$

For $p = 2$ we obtain the well-known *Euclidean norm* and the *Euclidean metric*.

In many practical applications generalizations of l_p norms, the so-called *weighted l_p norms*, are used to approximate travel distances (see Love and Morris, 1972, 1979, 1988):

Definition 1.4. *Let $w = (w_1, \ldots, w_n)^T \in \mathbb{R}^n$ with $w_i > 0$, $i = 1, \ldots, n$; let $p \in [1, \infty]$; and let $X = (x_1, \ldots, x_n)^T \in \mathbb{R}^n$. Then the weighted l_p norm of X is defined as*

$$l_p^w(X) = \left(\sum_{i=1}^{n} w_i |x_i|^p \right)^{\frac{1}{p}} \quad for\ p \in [1, \infty), \tag{1.14}$$

$$l_\infty^w(X) = \max_{i=1,\ldots,n} w_i |x_i|. \tag{1.15}$$

To illustrate the implications of the weights $w_1, \ldots, w_n$, the unit balls of the Euclidean norm and of the weighted Euclidean norm in $\mathbb{R}^2$ are shown for comparison in Figure 1.1.

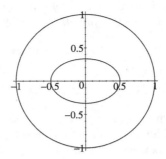

Figure 1.1. The unit balls of the Euclidean norm and of the weighted Euclidean norm with weights $w = (4,9)^T$ in $\mathbb{R}^2$.

The determination of suitable parameters w and p to obtain an accurate distance-predicting model is discussed, for example, in Brimberg and Love (1991, 1992), Brimberg et al. (1996), and Üster and Love (1998a,b).

For $p = 1$ and $p = \infty$, the unit ball of the l_p norm and of the weighted l_p norm is a polytope. The l_1 norm is often referred to as the *Manhattan norm* or *rectangular norm*. The l_∞ norm is also called the *Chebyshev norm* or *maximum norm*. In general, norms with a polyhedral unit ball are called *block norms*. Block norms and their generalization, the *polyhedral gauges*, will be discussed in more detail in the following section.

1.2 General Gauges and Polyhedral Distance Functions

One of the basic properties in the definition of norms is the symmetry of the unit ball. Symmetry implies, for example, that for any norm γ in $\mathbb{R}^n$,

$$\gamma(-X) = \gamma(X) \qquad \forall X \in \mathbb{R}^n.$$

However, distances in realistic environments may not always be symmetric. Consider, for example, two cities, one of which, city X, is located in a deep valley, whereas the other, city Y, is located on top of a high mountain next to the valley. Travel costs may be expected to be much higher for traveling from X to Y than vice versa.

If the symmetry assumption is dropped in the definition of a distance measure, we obtain the more general concept of *gauges*.

Definition 1.5. (Minkowski, 1911) *Let S be a compact and convex set in $\mathbb{R}^n$ containing the origin in its interior and let $X \in \mathbb{R}^n$. The* gauge *$\gamma : \mathbb{R}^n \to \mathbb{R}$ of X is defined as*

$$\gamma(X) := \inf\{\lambda > 0 : X \in \lambda S\}. \tag{1.16}$$

Similarly to (1.5) the *unit ball* of a gauge γ is given by

$$S = \{X \in \mathbb{R}^n \ : \ \gamma(X) \leq 1\}. \tag{1.17}$$

The *gauge distance* $\gamma(X, Y)$ between two points $X, Y \in \mathbb{R}^n$ is defined as

$$\gamma(X, Y) := \gamma(Y - X); \tag{1.18}$$

see Definition 1.2. Note that in general, $\gamma(X, Y) \neq \gamma(Y, X)$, since in Definition 1.5 a gauge γ is not assumed to be symmetric.

Lemma 1.3. (Minkowski, 1911) *Let γ be defined according to Definition 1.5. Then for all $X, Y \in \mathbb{R}^n$,*

$$\gamma(X) \geq 0 \quad and \quad (\gamma(X) = 0 \Leftrightarrow X = \underline{0}), \tag{1.19}$$
$$\gamma(\lambda X) = \lambda \gamma(X) \quad \forall \lambda \geq 0, \tag{1.20}$$
$$\gamma(X + Y) \leq \gamma(X) + \gamma(Y). \tag{1.21}$$

Note that, in the case that S is symmetric, γ defines a norm and (1.20) can be replaced by the stronger statement (1.3).

In the following we will concentrate on the special case that $S = P$ is a polytope. Distance functions based on polyhedral unit balls were introduced to location modeling in Ward and Wendell (1980, 1985). Since then they have been successfully applied to approximate road distances and other network distances.

Definition 1.6. (Ward and Wendell, 1985) *Let γ be defined according to Definition 1.5. If additionally the set S is a polytope $P = S$, then γ is called a* polyhedral gauge. *If P is also symmetric, γ is called a* block norm.

Let $\text{ext}(P) = \{v^1, \ldots, v^\delta\}$ *be the set of extreme points of P. Then $v^1, \ldots, v^\delta$ are called* fundamental vectors. *The half-lines $d^1, \ldots, d^\delta$ starting at the origin and passing through the extreme points $v^1, \ldots, v^\delta$, i.e., $d_i = \{\lambda v^i : \lambda \geq 0\}$, $i = 1, \ldots, \delta$, are called* fundamental directions.

Widmayer et al. (1987) introduced similar distance functions, called *distances in fixed orientations*, in the context of computational geometry. In this case it is assumed that P is symmetric and that all fundamental vectors have unit length; that is, $\|v^i\|_{l_2} = 1$, $i = 1, \ldots, \delta$.

In the following we will enumerate the fundamental vectors and the fundamental directions of a polyhedral gauge by superscripts and denote the components of the fundamental vectors and fundamental directions by subscripts.

Definition 1.7. (Ward and Wendell, 1985) *The fundamental vectors defined by the extreme points of the same facet of P span a* fundamental cone. *In $\mathbb{R}^2$, the fundamental cone spanned by two adjacent fundamental vectors v^i and v^j is referred to as $C\left(v^i, v^j\right)$.*

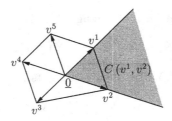

Figure 1.2. A polyhedral gauge with five fundamental vectors and the fundamental cone $C\left(v^1, v^2\right)$ spanned by v^1 and v^2.

An example of a polyhedral gauge and its fundamental vectors $v^1, \ldots, v^5$ in $\mathbb{R}^2$ is given in Figure 1.2.

Beyond the fact that polyhedral gauges and block norms can be used to approximate road distances, they were also proven to approximate any given gauge and any round norm arbitrarily well:

Theorem 1.1. (Ward and Wendell, 1985) *The set of polyhedral gauges is dense in the set of all gauges. Moreover, the set of block norms is dense in the set of all norms.*

Another advantageous property of polyhedral gauges is that they can be evaluated using the fundamental vectors only:

Lemma 1.4. (Ward and Wendell, 1985; Nickel, 1995) *Let P be a convex and compact polytope containing the origin in its interior, let* $\text{ext}(P) = \left\{v^1, \ldots, v^\delta\right\}$. *and let $X \in \mathbb{R}^n$. Then*

$$\gamma(X) = \min \left\{ \sum_{i=1}^{\delta} \lambda_i \; : \; X = \sum_{i=1}^{\delta} \lambda_i v^i \text{ and } \lambda_i \geq 0 \quad \forall i = 1, \ldots, \delta \right\}. \quad (1.22)$$

Lemma 1.4 indicates that the value of γ at a given point $X \in \mathbb{R}^n$ can be obtained by solving a linear programming problem for $\lambda_1, \ldots, \lambda_\delta$.

1.3 Polyhedral Gauges in $\mathbb{R}^2$

Since planar models are of particular interest in location modeling, we will concentrate in this section on the two-dimensional case and discuss some further properties of polyhedral gauges and block norms. Parts of the results presented in this section and in Section 1.4 were published in Dearing et al. (2002) and Hamacher and Klamroth (2000). A detailed survey on this subject can also be found in Schandl (1998).

In the following we consider a polyhedral gauge γ in $\mathbb{R}^2$ according to Definition 1.6 with unit ball P and fundamental vectors $\text{ext}(P) = \left\{v^1, \ldots, v^\delta\right\}$. Furthermore, we assume that $X \in \mathbb{R}^2$, $X \neq \underline{0}$, is an arbitrary point.

First observe that if $X \in C\left(v^i, v^j\right)$, where $C\left(v^i, v^j\right)$ is a fundamental cone, then the definition of convex cones implies that there exist unique scalars λ_i and λ_j such that $X = \lambda_i v^i + \lambda_j v^j$ and $\lambda_i, \lambda_j \geq 0$. Conversely, if v^i and v^j are the fundamental vectors that span the fundamental cone $C\left(v^i, v^j\right)$ and if $X = \lambda_i v^i + \lambda_j v^j$ is the unique representation of X in terms of v^i and v^j with $\lambda_i, \lambda_j \geq 0$, then $X \in C\left(v^i, v^j\right)$. This observation leads to the following representation of a polyhedral gauge γ; see also Widmayer et al. (1987) for a similar statement in the context of distances in fixed orientations:

Lemma 1.5. *Let γ be a polyhedral gauge in $\mathbb{R}^2$, let $X \in \mathbb{R}^2$ be in the fundamental cone $C\left(v^i, v^j\right)$, and let $X = \lambda_i v^i + \lambda_j v^j$ be the unique representation of X in terms of v^i and v^j. Then*

$$\gamma(X) = \lambda_i + \lambda_j. \tag{1.23}$$

Proof. First note that $\lambda_i, \lambda_j \geq 0$, since $X \in C\left(v^i, v^j\right)$.

Let Z be the intersection point of the boundary ∂P of P with the half-line starting at the origin and passing through X. Then $X = \gamma(X)Z$, where $Z = \alpha v^i + (1-\alpha)v^j$ for some $\alpha \in [0, 1]$; see Figure 1.3.

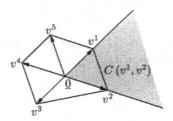

Figure 1.3. The points X and Z in $C\left(v^i, v^j\right)$.

We therefore obtain two different representations of X:

$$X = \lambda_i v^i + \lambda_j v^j,$$

$$X = \gamma(X)Z = \gamma(X)\alpha v^i + \gamma(X)(1-\alpha)v^j.$$

Combining these two representations yields

$$\lambda_i = \gamma(X)\alpha \quad \text{and} \quad \lambda_j = \gamma(X)(1-\alpha)$$

and thus $\gamma(X) = \lambda_i + \lambda_j$. $\qquad \square$

For a given point $X \in C\left(v^i, v^j\right)$, the value of $\gamma(X)$ can thus be determined using only the two fundamental vectors v^i and v^j in the following way (see Schandl, 1998):

Assume that the fundamental vectors of P are given in clockwise order and let $v^{\delta+1} := v^1$. Then $X \in C\left(v^i, v^{i+1}\right)$ for some $i \in \{1, \ldots, \delta\}$. Using the unique representation of X in terms of v^i and v^{i+1}, $X = \lambda_i v^i + \lambda_{i+1} v^{i+1}$, we can calculate that

$$\lambda_i = \frac{x_2 v_1^{i+1} - x_1 v_2^{i+1}}{v_1^{i+1} v_2^i - v_2^{i+1} v_1^i} \quad \text{and} \quad \lambda_{i+1} = \frac{x_1 v_2^i - x_2 v_1^i}{v_1^{i+1} v_2^i - v_2^{i+1} v_1^i}. \tag{1.24}$$

Since $\gamma(X) = \lambda_i + \lambda_{i+1}$, it follows that

$$\gamma(X) = \frac{x_1 \left(v_2^i - v_2^{i+1}\right) - x_2 \left(v_1^i - v_1^{i+1}\right)}{v_1^{i+1} v_2^i - v_2^{i+1} v_1^i}. \tag{1.25}$$

In the following we give some examples of polyhedral gauges in $\mathbb{R}^2$. In all of these examples, the unit balls of the corresponding polyhedral gauges are symmetric with respect to the origin, implying that they are in fact block norms.

Example 1.1. *The l_∞ norm and the weighted l_∞^w norm are defined as*

$$\|X\|_{l_\infty} = \max_{i=1,2} |x_i|,$$
$$\|X\|_{l_\infty}^w = \max_{i=1,2} w_i |x_i|,$$

respectively, where $w_1, w_2 > 0$ (see Figure 1.5). The fundamental vectors of the l_∞ norm in $\mathbb{R}^2$ are given by

$$v^1 = \begin{pmatrix} 1 \\ 1 \end{pmatrix}, \quad v^2 = \begin{pmatrix} 1 \\ -1 \end{pmatrix}, \quad v^3 = \begin{pmatrix} -1 \\ -1 \end{pmatrix}, \quad v^4 = \begin{pmatrix} -1 \\ 1 \end{pmatrix}.$$

For the weighted l_∞^w norm we obtain

$$v^1 = \begin{pmatrix} 1/w_1 \\ 1/w_2 \end{pmatrix}, \quad v^2 = \begin{pmatrix} 1/w_1 \\ -1/w_2 \end{pmatrix}, \quad v^3 = \begin{pmatrix} -1/w_1 \\ -1/w_2 \end{pmatrix}, \quad v^4 = \begin{pmatrix} -1/w_1 \\ 1/w_2 \end{pmatrix}.$$

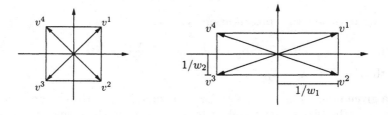

Figure 1.4. Unit balls of the l_∞ norm and of the weighted l_∞^w norm in $\mathbb{R}^2$.

Example 1.2. *The l_1 norm and the weighted l_1^w norm are defined as*

$$\|X\|_{l_1} = \sum_{i=1,2} |x_i|,$$

$$\|X\|_{l_1}^w = \sum_{i=1,2} w_i |x_i|,$$

respectively, where $w_1, w_2 > 0$ (see Figure 1.5). The fundamental vectors of the l_1 norm in $\mathbb{R}^2$ are given by

$$v^1 = \begin{pmatrix} 0 \\ 1 \end{pmatrix}, \quad v^2 = \begin{pmatrix} 1 \\ 0 \end{pmatrix}, \quad v^3 = \begin{pmatrix} 0 \\ -1 \end{pmatrix}, \quad v^4 = \begin{pmatrix} -1 \\ 0 \end{pmatrix},$$

and the fundamental vectors of the weighted l_1^w norm are given by

$$v^1 = \begin{pmatrix} 0 \\ 1/w_2 \end{pmatrix}, \quad v^2 = \begin{pmatrix} 1/w_1 \\ 0 \end{pmatrix}, \quad v^3 = \begin{pmatrix} 0 \\ -1/w_2 \end{pmatrix}, \quad v^4 = \begin{pmatrix} -1/w_1 \\ 0 \end{pmatrix}.$$

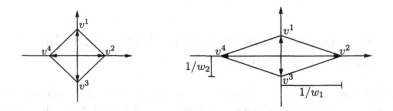

Figure 1.5. Unit balls of the l_1 norm and of the weighted l_1^w norm in $\mathbb{R}^2$.

Example 1.3. *The* weighted one infinity norm $l_{1,\infty}^{\alpha,\beta}$ *is defined as the sum of the weighted l_∞^α and l_1^β norms:*

$$\|X\|_{l_{1,\infty}}^{\alpha,\beta} = \|X\|_{l_1}^\alpha + \|X\|_{l_\infty}^\beta = \sum_{i=1,2} \alpha_i |x_i| + \max_{i=1,2} \beta_i |x_i|,$$

where $\alpha_1, \alpha_2, \beta_1, \beta_2 > 0$ (see Figure 1.6). The fundamental vectors of the weighted one infinity norm in $\mathbb{R}^2$ are as follows:

$$v^1 = \begin{pmatrix} 0 \\ \frac{1}{\beta_2 + \alpha_2} \end{pmatrix}, \qquad v^2 = \begin{pmatrix} \frac{1}{\beta_1 + \alpha_1 + \frac{\beta_1}{\beta_2}\alpha_2} \\ \frac{1}{\beta_2 + \alpha_2 + \frac{\beta_2}{\beta_1}\alpha_1} \end{pmatrix},$$

$$v^3 = \begin{pmatrix} \frac{1}{\beta_1 + \alpha_1} \\ 0 \end{pmatrix}, \qquad v^4 = \begin{pmatrix} \frac{1}{\beta_1 + \alpha_1 + \frac{\beta_1}{\beta_2}\alpha_2} \\ -\frac{1}{\beta_2 + \alpha_2 + \frac{\beta_2}{\beta_1}\alpha_1} \end{pmatrix},$$

$$v^5 = \begin{pmatrix} 0 \\ -\frac{1}{\beta_2 + \alpha_2} \end{pmatrix}, \qquad v^6 = \begin{pmatrix} -\frac{1}{\beta_1 + \alpha_1 + \frac{\beta_1}{\beta_2}\alpha_2} \\ -\frac{1}{\beta_2 + \alpha_2 + \frac{\beta_2}{\beta_1}\alpha_1} \end{pmatrix},$$

$$v^7 = \begin{pmatrix} -\frac{1}{\beta_1 + \alpha_1} \\ 0 \end{pmatrix}, \qquad v^8 = \begin{pmatrix} -\frac{1}{\beta_1 + \alpha_1 + \frac{\beta_1}{\beta_2}\alpha_2} \\ \frac{1}{\beta_2 + \alpha_2 + \frac{\beta_2}{\beta_1}\alpha_1} \end{pmatrix}.$$

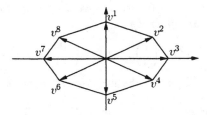

Figure 1.6. Unit ball of the weighted one infinity norm in $\mathbb{R}^2$.

1.4 Relations Between Block Norms and the Manhattan Norm

In the following we will focus on block norms in the 2-dimensional real space $\mathbb{R}^2$ with exactly four fundamental vectors (see Figure 1.7 for an example) and discuss their relation to the Manhattan norm l_1 (see Figure 1.5). To distinguish these more general distance functions from the Manhattan norm with the fundamental vectors $v^1, \ldots, v^4$ as defined in Example 1.2, we will in the following denote the fundamental vectors of any other block norm with four fundamental vectors by the vectors $u^1, \ldots, u^4$.

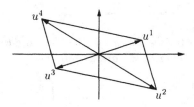

Figure 1.7. The unit ball of a block norm with four fundamental vectors $u^1, \ldots, u^4$.

Let γ be a block norm with the four fundamental vectors (in clockwise order)

$$u^1 = \left(u_1^1, u_2^1\right)^T, \quad u^2 = \left(u_1^2, u_2^2\right)^T, \quad u^3 = \left(u_1^3, u_2^3\right)^T, \quad u^4 = \left(u_1^4, u_2^4\right)^T$$

such that $u^3 = -u^1$ and $u^4 = -u^2$. If T is the linear transformation such that $T\left(u^1\right) = Tu^1 = v^1$ and $T\left(u^2\right) = Tu^2 = v^2$, then

$$T = \begin{pmatrix} t_{11} & t_{12} \\ t_{21} & t_{22} \end{pmatrix} = \frac{1}{u_2^1 u_1^2 - u_1^1 u_2^2} \begin{pmatrix} u_2^1 & -u_1^1 \\ -u_2^2 & u_1^2 \end{pmatrix}. \tag{1.26}$$

Observe that the corresponding inverse transformation is given by

$$T^{-1} = \begin{pmatrix} u_1^2 & u_1^1 \\ u_2^2 & u_2^1 \end{pmatrix}.$$

Setting $T(X) := TX$ for all $X \in \mathbb{R}^2$, we can prove the following result relating block norm distances to the l_1 metric:

Lemma 1.6. *Let γ be a block norm with four fundamental vectors and let the linear transformation T be defined by (1.26). Then*

$$\gamma(X, Y) = l_1(T(X), T(Y))$$

for all $X, Y \in \mathbb{R}^2$.

Proof. First consider the special case that $Y = \underline{0}$. If also $X = \underline{0}$, the result is trivial. Let $X \in C\left(u^i, u^{i+1}\right)$ for some $i \in \{1, \ldots, 4\}$ and let $X = \lambda_i u^i + \lambda_{i+1} u^{i+1}$ be the unique representation of X in terms of u^i and u^{i+1} (see Lemma 1.5).

Using the fact that γ has only four fundamental vectors we can assume without loss of generality that $u^i \in \{u^1, -u^1\}$ and $u^{i+1} \in \{u^2, -u^2\}$. Using (1.24) we can calculate

$$\begin{aligned} l_1(T(X), \underline{0}) &= |t_{11}x_1 + t_{12}x_2| + |t_{21}x_1 + t_{22}x_2| \\ &= \left| \frac{u_2^1 x_1 - u_1^1 x_2}{u_2^1 u_1^2 - u_1^1 u_2^2} \right| + \left| \frac{u_1^2 x_2 - u_2^2 x_1}{u_2^1 u_1^2 - u_1^1 u_2^2} \right| \\ &= \lambda_{i+1} + \lambda_i \\ &= \gamma(X, \underline{0}). \end{aligned}$$

Now consider the general case that $X, Y \in \mathbb{R}^2$ are both different from $\underline{0}$. Since T is a linear transformation, we immediately obtain

$$\begin{aligned} l_1(T(X), T(Y)) &= l_1(T(X) - T(Y), \underline{0}) = l_1(T(X - Y), \underline{0}) \\ &= \gamma(X - Y, \underline{0}) = \gamma(X, Y). \end{aligned} \qquad \square$$

Note that this result is well known for the special case of the Chebyshev metric l_∞. In this case the transformation T is given by

$$T = \frac{1}{2} \begin{pmatrix} 1 & -1 \\ 1 & 1 \end{pmatrix};$$

see, for example, Francis et al. (1992), Hamacher (1995). Note also that the definition of T for a given block norm γ depends on the assignment of $T\left(u^1\right) = v^1$ and changes if we set $T\left(u^1\right) = -v^1$, $T\left(u^1\right) = v^2$ or $T\left(u^1\right) = -v^2$.

2
Shortest Paths in the Presence of Barriers

The problem of finding shortest paths in realistic environments plays an important role not only in location planning. Shortest-path problems in the presence of physical barriers arise, for example, in the planning of shortest water routes between different harbors or in the determination of an optimal path of a robot in an industrial plant.

Perhaps the first example of an application of a motion planning problem in the presence of barriers was addressed in Kirk and Lim (1970). The authors studied the problem of finding optimal routes for an autonomous vehicle exploring the surface of Mars. Four years later, Wangdahl et al. (1974) discussed the optimal layout for a pipeline connecting two points on a ship. Vaccaro (1974) considered routing problems in congested areas, and Lozano-Perez and Wesley (1979) worked on motion-planning problems for robots in the presence of polyhedral obstacles.

More recently, the importance of motion-planning problems especially in the context of robot control is reflected in an increasing number of scientific publications in this field. For an overview of recent developments, algorithms, and applications we refer to the textbook of Latombe (1991), the survey chapter of Mitchell (2000), and the Ph.D. thesis of Verbarg (1996). Further references can be found, for example, in Alt and Yap (1990), Schwartz and Sharir (1990), Schwartz and Yap (1987), and Van Der Stappen (1994).

Due to the variety of applications, a path in the presence of barriers can be an *optimal path* with respect to different criteria. A natural definition of an optimal path is that of a *shortest path*, that is, a minimum-length path with respect to some given distance measure. Other definitions include that of *minimum-link paths* (Mitchell et al., 1992; Suri, 1986), motivated by robots moving only on straight line segments and inducing high costs for every change of direction. In this case the objective is to minimize the number of turns of a path consisting of a finite number of straight line segments. Especially in environments with high security standards the notion of a *safest path* (Roos and Noltemeier, 1991) is needed, i.e., a path maximizing the safety distance to a set of obstacles.

Since our main interest will be the optimal location of one or several facilities that minimize some kind of transportation cost, we will focus our attention on the first definition of optimal paths as *shortest paths*.

2.1   Shortest Paths and the Concept of Visibility

In the following we will introduce the basic definitions related to shortest paths in the presence of barriers. The results presented in this section were also published partially in Dearing et al. (2002) and Klamroth (2001b).

Let a prescribed metric d be given in the n-dimensional space $\mathbb{R}^n$ ($n \geq 2$) measuring travel distances between pairs of points in $\mathbb{R}^n$. We assume that the metric d is induced by a norm $\|\bullet\|_d : \mathbb{R}^n \to \mathbb{R}$ and is therefore symmetric (1.8) and satisfies the triangle inequality (1.9). Thus d can be calculated with respect to $\|\bullet\|_d$ as

$$d(X, Y) = \|Y - X\|_d \qquad \forall\, X, Y \in \mathbb{R}^n.$$

Additionally, let $\{B_1, \ldots, B_N\}$ be a finite set of closed and pairwise disjoint sets in $\mathbb{R}^n$ with nonempty interior. Each set B_i, $i = 1, \ldots, N$, is called a *barrier* (or *obstacle*), and the union of all barriers is denoted by $\mathcal{B}$; i.e., $\mathcal{B} = \bigcup_{i=1}^{N} B_i$. Note that we do not assume each barrier set to be bounded at this point. We assume that traveling is prohibited in the interior of the barriers. However, traveling is allowed along the boundary $\partial(B_i)$ of each barrier region B_i, $i = 1, \ldots, N$.

Let $\mathcal{F} := \mathbb{R}^n \setminus \text{int}(\mathcal{B})$ be the *feasible region* (*free space*) in $\mathbb{R}^n$. To avoid infeasibility we assume that $\mathcal{F}$ is a connected subset of $\mathbb{R}^n$.

The *barrier distance function* (also referred to as *shortest path metric* or *geodesic distance*) $d_\mathcal{B} : \mathcal{F} \times \mathcal{F} \to \mathbb{R}$ then measures the length of a shortest path between two points X and Y in the feasible region $\mathcal{F} = \mathbb{R}^n \setminus \text{int}(\mathcal{B})$ not intersecting the interior of a barrier:

Definition 2.1. *A continuous curve P given by the parameterization $p = (p_1, \ldots, p_n)^T : [0, 1] \to \mathbb{R}^n$ with $p(0) = X$ and $p(1) = Y$ that is continuously differentiable on $[0, 1]$ with the possible exception of at most a finite number of points, where the derivative $p' = (p'_1, \ldots, p'_n)^T$ has finite limits from the left and from the right, is called an X-Y path.*

The length $l(P)$ of the X-Y path P with respect to the prescribed metric d is given by

$$l(P) := \int_0^1 \|p'(t)\|_d dt.$$

If P does not intersect the interior of a barrier, i.e., if $p([0,1]) \cap \text{int}(\mathcal{B}) = \emptyset$, the X-Y path P is called a permitted *X-Y path.*

Using the notion of a permitted X-Y path, the *barrier distance $d_{\mathcal{B}}$* between two points X and Y in the feasible region $\mathcal{F}$ can be defined as follows:

Definition 2.2. *For $X, Y \in \mathcal{F}$ the* barrier distance $d_{\mathcal{B}}(X, Y)$ *is defined as the infimum of the lengths of all permitted X-Y paths; i.e.,*

$$d_{\mathcal{B}}(X, Y) := \inf \{l(P) \, : \, P \text{ permitted } X\text{-}Y \text{ path}\}. \qquad (2.1)$$

A permitted X-Y path with length $d_{\mathcal{B}}(X,Y)$ is called a d-shortest permitted X-Y path (or optimal X-Y path, geodesic path).

Note that the existence of a d-shortest permitted X-Y path that is a smooth curve (with the possible exception of at most a finite number of points) cannot always be guaranteed. In particular, there might exist cases where the infimum in (2.1) is attained for a path not satisfying the differentiability requirements of Definition 2.1. An example for the 2-dimensional case is given in Figure 2.1.

In the following we will mostly concentrate on problems (and barrier sets) for which a d-shortest permitted X-Y path exists for all choices of $X, Y \in$

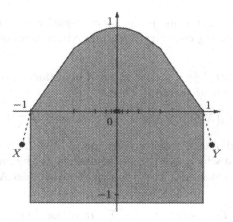

Figure 2.1. An example in $\mathbb{R}^2$ where no l_2-shortest permitted X-Y path exists. The upper boundary of the barrier set B has an infinite number of extreme points at $(1/i, 1 - 1/i^2)^T$ and $(-1/i, 1 - 1/i^2)^T$, $i \in \{2, 4, 8, 16, \dots\}$, and therefore a path along this boundary is not a permitted X-Y path according to Definition 2.1.

$\mathcal{F}$. This is, for example, the case if a finite number of smooth or polyhedral barrier sets is given in $\mathbb{R}^n$. In these cases, the infimum in (2.1) can be replaced by a minimum.

The barrier distance $d_{\mathcal{B}}$ is in general not positively homogeneous, which implies that there does not exist a norm inducing the metric $d_{\mathcal{B}}$. Moreover, the barrier distance $d_{\mathcal{B}}$ is in general nonconvex. Nevertheless, $d_{\mathcal{B}}$ defines a metric on the feasible region $\mathcal{F}$:

Lemma 2.1. *Let d be a metric induced by a norm and let $\mathcal{B} = \bigcup_{i=1}^{N} B_i$ be the union of a set of barriers. Then $d_{\mathcal{B}}$ defines a metric on $\mathcal{F}$; i.e.,*

$$d_{\mathcal{B}}(X,Y) \geq 0 \quad and \quad (d_{\mathcal{B}}(X,Y) = 0 \Leftrightarrow X = Y) \quad \forall X, Y \in \mathcal{F}, \qquad (2.2)$$

$$d_{\mathcal{B}}(X,Y) = d_{\mathcal{B}}(Y,X) \qquad\qquad\qquad \forall X, Y \in \mathcal{F}, \qquad (2.3)$$

$$d_{\mathcal{B}}(X,Y) \leq d_{\mathcal{B}}(X,Z) + d_{\mathcal{B}}(Z,Y) \qquad\qquad \forall X, Y, Z \in \mathcal{F}. \qquad (2.4)$$

Consequently, $(\mathcal{F}, d_{\mathcal{B}})$ is an n-dimensional metric space.

Proof. Equations (2.2) and (2.3) are trivial. For the triangle inequality (2.4) consider three points $X, Y, Z \in \mathcal{F}$. Then for every permitted X-Z path $P_{X,Z}$ and for every permitted Z-Y path $P_{Z,Y}$ there exists a permitted X-Y path $P_{X,Y} = P_{X,Z} \cup P_{Z,Y}$ of length $l(P_{X,Y}) = l(P_{X,Z}) + l(P_{Z,Y})$, passing through Z. Thus,

$$
\begin{aligned}
d_{\mathcal{B}}(X,Y) &= \inf\{l(P_{X,Y}) : P_{X,Y} \text{ permitted } X\text{-}Y \text{ path}\} \\
&\leq \inf\{l(P_{X,Z}) + l(P_{Z,Y}) : P_{X,Z} \text{ permitted } X\text{-}Z \text{ path and} \\
&\qquad\qquad\qquad\qquad\qquad P_{Z,Y} \text{ permitted } Z\text{-}Y \text{ path}\} \\
&= d_{\mathcal{B}}(X,Z) + d_{\mathcal{B}}(Z,Y). \qquad\qquad\qquad\qquad\qquad \square
\end{aligned}
$$

We will refer to the metric $d_{\mathcal{B}}$ as the *shortest-path metric* or as the *geodesic distance*. The following corollary is an immediate consequence of the definition of $d_{\mathcal{B}}$:

Corollary 2.1. *Let d be a metric induced by a norm and let $\mathcal{B} = \bigcup_{i=1}^{N} B_i$ be the union of a set of barriers. Then*

$$d_{\mathcal{B}}(X,Y) \geq d(X,Y) \qquad\qquad \forall X, Y \in \mathcal{F}.$$

Considering the distance from a given point $X \in \mathcal{F}$ we can distinguish those parts of $\mathcal{F}$ in which $d_{\mathcal{B}}(X,Y)$ is equivalent to the metric $d(X,Y)$ and those parts of $\mathcal{F}$ where $d_{\mathcal{B}}(X,Y) > d(X,Y)$ (see also Aneja and Parlar, 1994):

Definition 2.3. *Two points $X, Y \in \mathcal{F}$ are called d-visible if*

$$d_{\mathcal{B}}(X,Y) = d(X,Y),$$

that is, if the distance between X and Y is not lengthened by the barrier regions.

The set of points that are d-visible from a point $X \in \mathcal{F}$ is defined as

$$\text{visible}_d(X) := \{Y \in \mathcal{F} \ : \ d_{\mathcal{B}}(X, Y) = d(X, Y)\}.$$

Similarly, we call the set of points $Y \in \mathcal{F}$ that are not d-visible from a point $X \in \mathcal{F}$ the d-shadow of X; i.e.,

$$\text{shadow}_d(X) := \{Y \in \mathcal{F} \ : \ d_{\mathcal{B}}(X, Y) > d(X, Y)\}.$$

Definition 2.4. *The boundary of the closure of the d-shadow of a point $X \in \mathcal{F}$ (or, for brevity, the boundary of $\text{shadow}_d(X)$) is defined as*

$$\partial(\text{shadow}_d(X)) \ := \ \{Y \in \mathcal{F} \ : \ N_\varepsilon(Y) \cap \text{shadow}_d(Y) \neq \emptyset$$
$$\text{and } N_\varepsilon(Y) \not\subseteq \text{shadow}_d(Y) \ \forall \varepsilon > 0\},$$

where $N_\varepsilon(Y) := \{Z \in \mathbb{R}^n : l_2(Z, Y) < \varepsilon\}$.

Two examples for possible shapes of the d-shadow of a given point $X \in \mathbb{R}^2$ with respect to two different metrics d are given in Figure 2.2.

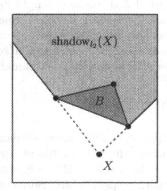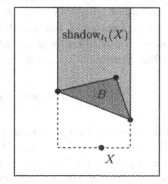

Figure 2.2. The l_2-shadow and the l_1-shadow of a point $X \in \mathbb{R}^2$.

Observe that for some choices of the metric d a point that is d-visible may not be l_2-visible, i.e., not visible in the usual sense of straight line visibility. On the other hand, every pair of l_2-visible points is also d-visible if d is a metric induced by a norm, as the following lemma shows.

Lemma 2.2. *Let d be a metric induced by a norm. Then*

$$\text{shadow}_d(X) \subseteq \text{shadow}_{l_2}(X), \qquad X \in \mathcal{F}.$$

Furthermore, if $X, Y \in \mathcal{F}$ are l_2-visible, $X \neq Y$, then the straight line segment connecting X and Y is a d-shortest permitted X-Y path with length $d(X, Y)$.

Proof. Without loss of generality let $X = \underline{0}$ be the origin and let $Y \in \mathcal{F}$ be a point that is l_2-visible from X. Then the straight line segment connecting X and Y is a permitted X-Y path P given by the parameterization p : $[0, 1] \to \mathbb{R}^n$, $p(t) = t \cdot Y$, $t \in [0, 1]$. Using the definition of the barrier distance $d_\mathcal{B}$, Definition 2.2, the length of P can be calculated as

$$d_\mathcal{B}(\underline{0}, Y) \leq l(P) = \int_0^1 \|p'(t)\|_d dt = \int_0^1 \left\| \frac{d}{dt}(tY) \right\|_d dt = \int_0^1 \|Y\|_d dt$$

$$= \|Y\|_d = d(\underline{0}, Y).$$

Thus the straight line segment connecting X and Y is a d-shortest permitted X-Y path with length $d(X, Y)$. Consequently, the subregion of $\mathcal{F}$ that is l_2-visible from a point $X \in \mathcal{F}$ is also d-visible from X, and $\text{shadow}_d(X) \subseteq \text{shadow}_{l_2}(X)$ for all $X \in \mathcal{F}$. $\square$

2.2 Optimality Conditions for Smooth Barriers

The problem of finding d-shortest permitted X-Y paths of length $d_\mathcal{B}(X, Y)$ between two given points $X, Y \in \mathcal{F}$ can, for certain distance functions d and for certain "smooth" barrier shapes, be interpreted as a special case of a problem in the calculus of variations (see, for example, Clarke, 1983; Elsgolc, 1962; Smith, 1974). In its general formulation a functional

$$J = J(p) = \int_0^1 F(t, p(t), p'(t)) dt$$

is to be minimized over all piecewise continuously differentiable functions p : $[0, 1] \to \mathbb{R}^n$, $p(t) = (p_1(t), \dots, p_n(t))^T$ under the fixed endpoint conditions $p(0) = X$ and $p(1) = Y$. The function $F : [0, 1] \times \mathbb{R}^n \times \mathbb{R}^n \to \mathbb{R}$ is assumed to be twice continuously differentiable with respect to all arguments.

In our case where we want to find d-shortest permitted X-Y paths in $\mathbb{R}^n$, the function F is given by

$$F(t, p(t), p'(t)) := \|p'(t)\|_d,$$

and we wish to minimize the functional $J(p) := \int_0^1 \|p'(t)\|_d dt$ over all parameterizations p of permitted X-Y paths in $\mathcal{F}$.

Due to the assumptions on F we assume in the following that $\| \bullet \|_d$ is twice continuously differentiable. Examples include the l_p norms, $1 < p < \infty$, and thus in particular the Euclidean norm l_2.

For the planar case ($n = 2$), Smith (1974) showed that a d-shortest permitted X-Y path is always a composite curve consisting of pieces of arcs along which the Euler Lagrange differential equations $F_{p_i}(t, p, p') - \frac{d}{dt} F_{p_i'}(t, p, p') = 0$, $i = 1, 2$, hold, and pieces of arcs (or isolated points)

along the boundary of a barrier. In the Euclidean case this implies that
an optimal path consists of line segments and of pieces of arcs (or isolated
points) along the boundary of $\mathcal{B}$. In the following we will derive similar
conditions for the n-dimensional case.

For simplicity we assume that only one compact and convex barrier region $\mathcal{B} = B$ is given in $\mathbb{R}^n$.

The assumption on the boundedness of the barrier region simplifies the
derivation of an alternative representation of permitted paths in $\mathbb{R}^n$. This
representation will be used later in this section to develop necessary optimality conditions for d-shortest permitted paths in $\mathbb{R}^n$. Since in the case
that an unbounded barrier B is given, a (possibly very large) bounded
barrier contained in B can be used without changing the barrier distance
between two given points $X, Y \in \mathcal{F}$, this assumption does not restrict the
generality of the model.

Moreover, the restriction to only one barrier region can be imposed without loss of generality, since an optimal path can always be decomposed such
that the problem can be formulated on several subregions of the $\mathbb{R}^n$ each
of which contains only one barrier region.

Additionally, we assume that the boundary $\partial(B)$ of B is given by an
$(n-1)$-dimensional smooth manifold in $\mathbb{R}^n$, which for all $X \in \partial(B)$ can
be represented relative to an open neighborhood $O \subseteq \mathbb{R}^n$ of X as the set
of solutions of $G_X(X) = 0$, where $G_X : O \to \mathbb{R}$ is a twice continuously
differentiable mapping with $\nabla G_X(X) \neq \underline{0}$. To avoid lengthy discussions
involving different representations of $\partial(B)$ we assume furthermore that
$\partial(B)$ can be represented with respect to only one functional $G : \mathbb{R}^n \to \mathbb{R}$
satisfying the above requirements as

$$\partial(B) = \{X \in \mathbb{R}^n \ : \ G(X) = 0\}. \qquad (2.5)$$

This assumption is used for simplicity only, while it is not essential for the
development of optimality conditions.

Under these assumptions, the problem of finding a d-shortest permitted X-Y path is equivalent to the following minimization problem over
all feasible and piecewise continuously differentiable X-Y paths P with
parameterization p (cf. Definition 2.1):

$$\begin{aligned}
\min \quad & J(p) = \int_0^1 F(t, p(t), p'(t))dt = \int_0^1 \|p'(t)\|_d dt \\
\text{s.t.} \quad & p(0) = X, \ p(1) = Y, \\
& p(t) \notin \text{int}(B) \qquad \forall t \in [0,1].
\end{aligned} \qquad (2.6)$$

This problem is a problem in the calculus of variations with the fixed endpoint conditions $p(0) = X$ and $p(1) = Y$ and with the additional constraint
$p(t) \notin \text{int}(B)$, $t \in [0,1]$.

Unfortunately, the Euler Lagrange theorem (see Smith, 1974) cannot be
applied immediately to (2.6) to derive necessary conditions for an optimal

solution p, since the feasible set, the set of all parameterizations p of permitted X-Y paths, is not an open set in the set of all piecewise continuously differentiable paths connecting X and Y. Moreover, we have to expect a similar situation as in the 2-dimensional case, where optimal paths are composite curves, which on some segments coincide with the boundary $\partial(B)$ of the barrier region B and thus do not allow variations in all directions on these segments.

In order to overcome this difficulty we introduce, for every parameterization p of a permitted X-Y path, a slack function $\mu : [0,1] \to \mathbb{R}$ combined with a vector-valued mapping $u : [0,1] \to \partial(B)$ on the boundary of B, defined by the relation

$$
\begin{aligned}
p(t) &= u(t) + \mu^2(t)N(u(t)). && u(t) \in \partial(B), \ \mu(t) \in \mathbb{R}, \\
\text{s.t.} \ \ G(u(t)) &= 0 && \forall t \in [0,1]
\end{aligned}
\tag{2.7}
$$

see Figure 2.3. Here $N(u) := \nabla G(u)$ denotes a particular (and unique) normal vector of $\partial(\mathcal{B})$ at the point $u = u(t) \in \partial(B)$ (see Rockafellar and Wets, 1998).

This representation is well-defined if the boundary $\partial(B)$ of the barrier set B has a representation (2.5) in an open neighborhood of $u(t) \in \partial(B)$ for all $t \in [0,1]$. Moreover, u and μ are continuous everywhere on $[0,1]$ and piecewise continuously differentiable with the possible exception of a finite number of points where the derivatives $u'(t)$ and $\mu'(t)$ have well-defined limiting values both from the left and from the right.

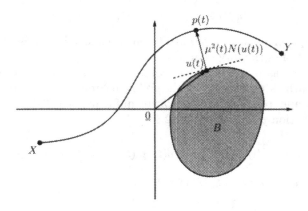

Figure 2.3. Representation of a point $p(t)$ on a permitted X-Y path in $\mathbb{R}^2$ with respect to $u(t)$ and $\mu(t)$, $t \in [0,1]$.

Example 2.1. *Let B be the unit sphere in $\mathbb{R}^n$, centered at the origin, with*

$$\partial(B) = \left\{ X = (x_1, \ldots, x_n)^T \in \mathbb{R}^n \ : \ \left(\sum_{i=1}^{n} x_i^2 \right)^{\frac{1}{2}} = 1 \right\}.$$

The boundary $\partial(B)$ of B can be represented in the form (2.5) by the functional $G : \mathbb{R}^n \to \mathbb{R}$ given by

$$G(X) = \left(\sum_{i=1}^{n} x_i^2 \right)^{\frac{1}{2}} - 1.$$

Let $X, Y \in \mathbb{R}^n$, $X, Y \notin \text{int}(B)$, be two feasible points. Then the parameterization p of any permitted X-Y path P can be written as

$$p(t) = u(t) + \mu^2(t) N(u(t)), \qquad u(t) \in \partial(B), \ \mu(t) \in \mathbb{R},$$
$$\text{s.t. } G(u(t)) = 0 \qquad\qquad \forall t \in [0, 1],$$

see Figure 2.4, where for all $\bar{u} \in \partial(B)$ the normal vector $N(\bar{u})$ is given by

$$N(\bar{u}) = \nabla G(\bar{u}) = \left(\frac{\partial}{\partial u_1} G(u), \ldots, \frac{\partial}{\partial u_n} G(u) \right) \Bigg|_{u = \bar{u}} = \bar{u}.$$

This can be easily verified, since the partial derivatives of G at a point $u = (u_1, \ldots, u_n)^T \in \partial(B)$ can be evaluated as

$$\frac{\partial}{\partial u_j} G(u) = \frac{1}{2} \left(\sum_{i=1}^{n} u_i^2 \right)^{-\frac{1}{2}} \cdot 2u_j, \qquad j = 1, \ldots, n,$$

and therefore

$$\frac{\partial}{\partial u_j} G(u) \Bigg|_{\substack{u = \bar{u}, \\ G(\bar{u}) = 0}} = \bar{u}_j, \qquad j = 1, \ldots, n,$$

for all $\bar{u} \in \mathbb{R}^n$ satisfying $G(\bar{u}) = 0$.

Given a point $p(t) \in \mathcal{F}$, $t \in [0, 1]$, on a permitted X-Y path P, we can conclude that $p(t) = \left(1 + \mu^2(t)\right) u(t)$ with $G(u(t)) = 0$ and therefore

$$\mu^2(t) = \|p(t)\|_{l_2} - 1,$$
$$u(t) = N(u(t)) = \frac{p(t)}{\|p(t)\|_{l_2}},$$

where $\|p(t)\|_{l_2} \geq 1$, since we assumed that $p(t) \notin \text{int}(B)$ for all $t \in [0, 1]$.

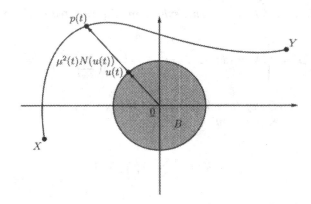

Figure 2.4. Representation of a point $p(t)$ with respect to $u(t)$ and $\mu(t)$, $t \in [0,1]$, in case of a circular barrier in $\mathbb{R}^2$.

Clearly, every pair of continuous and piecewise continuously differentiable functions $u : [0,1] \to \mathbb{R}^n$ and $\mu : [0,1] \to \mathbb{R}$ that satisfy the constraints

$$G(u(t)) = 0 \qquad \forall t \in [0,1],$$
$$u(0) + \mu^2(0)N(u(0)) = X,$$
$$u(1) + \mu^2(1)N(u(1)) = Y.$$

defines a unique X-Y path P such that $p(t) \cap \text{int}(B) = \emptyset$. Thus the original problem (2.6) of minimizing the functional $J(p)$ over all permitted X-Y paths P can be reformulated as the problem of minimizing the functional $\tilde{J}(u, \mu)$ given by

$$\tilde{J}(u, \mu) := \int_0^1 \tilde{F}(t, [u(t), \mu(t)], [u'(t), \mu'(t)])dt \qquad (2.8)$$

subject to the fixed endpoint conditions

$$u(0) + \mu^2(0)N(u(0)) = X \qquad \text{and} \qquad u(1) + \mu^2(1)N(u(1)) = Y \quad (2.9)$$

and the constraint

$$G(u(t)) = 0 \qquad \forall t \in [0,1]. \qquad (2.10)$$

Here $u : [0,1] \to \partial(B)$ and $\mu : [0,1] \to \mathbb{R}$ are both assumed to be continuous everywhere on $[0,1]$ and continuously differentiable on $[0,1]$ with the possible exception of at most a finite number of points where the derivatives $u'(t)$ and $\mu'(t)$ have well-defined limiting values both from the left and from the right.

The function $\tilde{F}$ in (2.8) is defined as

$$\tilde{F}(t, [u, \mu], [u', \mu']) := F\left(t, u(t) + \mu^2(t)N(u(t)), \frac{d}{dt}\left[u(t) + \mu^2(t)N(u(t))\right]\right).$$

Note that the set of solutions $[u, \mu]$ of (2.8) is an open set on which only the fixed endpoint conditions (2.9) and the equality constraint (2.10) are imposed. Therefore, the Euler Lagrange multiplier theorem can be applied to this reformulation of the problem. We can conclude that all local extremum functions $[u, \mu]$ must satisfy the following system of Euler Lagrange differential equations (see, for example, Elsgolc, 1962):

$$L_{u_i} - \frac{d}{dt} L_{u'_i} = 0, \qquad i = 1, \ldots, n, \qquad (2.11)$$

$$L_\mu - \frac{d}{dt} L_{\mu'} = 0, \qquad (2.12)$$

where L is a Lagrange function of the form

$$L(t, [u(t), \mu(t)], [u'(t), \mu'(t)]) = \tilde{F}(t, [u(t), \mu(t)], [u'(t), \mu'(t)]) + \lambda(t) G(u(t))$$

with Lagrange multipliers $\lambda(t) \in \mathbb{R}$, $t \in [0, 1]$, and where L_{u_i} and L_μ denote the partial derivatives of the Lagrange function L with respect to u_i and μ, respectively.

The solutions $[u, \mu]$ of the Euler Lagrange differential equations (2.11) and (2.12) and the Lagrange multipliers $\lambda(t)$ have to be determined such that the fixed endpoint conditions (2.9) and the constraint (2.10) are satisfied.

In order to further evaluate (2.11) and (2.12) we define for $i = 1, \ldots, n$ the substitutions

$$y_i := u_i(t) + \mu^2(t) N_i(u(t)) \qquad (2.13)$$

and

$$z_i := u'_i(t) + 2\mu(t)\mu'(t) N_i(u(t)) + \mu^2(t) \frac{d}{dt}(N_i(u(t))). \qquad (2.14)$$

Using this substitution and applying the chain rule of differential calculus we obtain

$$L_{u_i} = \sum_{j=1}^{n} \left[\mu^2 \frac{\partial N_j(u)}{\partial u_i} F_{p_j}(t, p, p') \right] + F_{p_i}(t, p, p')$$

$$+ \sum_{j=1}^{n} \left[\left(2\mu\mu' \frac{\partial N_j(u)}{\partial u_i} + \mu^2 \frac{\partial(\frac{d}{dt} N_j(u))}{\partial u_i} \right) F_{p'_j}(t, p, p') \right] + \lambda(t) G_{u_i}(u),$$

$$L_{u'_i} = \sum_{j=1}^{n} \left[\mu^2 \frac{\partial(\frac{d}{dt} N_j(u))}{\partial u'_i} F_{p'_j}(t, p, p') \right] + F_{p'_i}(t, p, p').$$

Observe that $\frac{\partial u_i'}{\partial u_i} = \frac{\partial}{\partial u_i}(\frac{d}{dt}u_i) = \frac{d}{dt}(\frac{\partial}{\partial u_i}u_i) = \frac{d}{dt}(1) = 0$. We can conclude that

$$\frac{d}{dt}L_{u_i'} = \sum_{j=1}^{n}\left[\left(2\mu\mu'\frac{\partial(\frac{d}{dt}N_j(u))}{\partial u_i'} + \mu^2\frac{d}{dt}\left(\frac{\partial(\frac{d}{dt}N_j(u))}{\partial u_i'}\right)\right)F_{p_j'}(t,p,p')\right]$$
$$+\sum_{j=1}^{n}\left[\mu^2\frac{\partial(\frac{d}{dt}N_j(u))}{\partial u_i'}\frac{d}{dt}F_{p_j'}(t,p,p')\right] + \frac{d}{dt}F_{p_i'}(t,p,p').$$

If we additionally use the fact that $\frac{\partial}{\partial u_i}\left(\frac{d}{dt}N_j(u)\right) = \frac{d}{dt}\left(\frac{\partial}{\partial u_i}N_j(u)\right)$, $i,j \in \{1,\ldots,n\}$, and that

$$\frac{\partial}{\partial u_i'}\left(\frac{d}{dt}N_j(u)\right) = \frac{\partial}{\partial u_i'}\left(\sum_{k=1}^{n}\left(\frac{\partial}{\partial u_k}N_j(u)\right)u_k'\right) = \frac{\partial N_j(u)}{\partial u_i}; \quad i,j \in \{1,\ldots,n\},$$

then the Euler Lagrange differential equations (2.11) can be rewritten as

$$\mu^2\sum_{j=1}^{n}\left[\frac{\partial N_j(u)}{\partial u_i}\left(F_{p_j}(t,p,p') - \frac{d}{dt}F_{p_j'}(t,p,p')\right)\right]$$
$$+ F_{p_i}(t,p,p') - \frac{d}{dt}F_{p_i'}(t,p,p') + \lambda(t)G_{u_i}(u) = 0, \qquad i = 1,\ldots,n.$$

Analogously, using the same substitutions (2.13) and (2.14) we can evaluate L_μ and $L_{\mu'}$ as

$$L_\mu = \sum_{j=1}^{n}\left[2\mu N_j(u)F_{p_j}(t,p,p')\right]$$
$$+ \sum_{j=1}^{n}\left[\left(2\mu'N_j(u) + 2\mu(\frac{d}{dt}N_j(u))\right)F_{p_j'}(t,p,p')\right],$$

$$L_{\mu'} = \sum_{j=1}^{n}\left[2\mu N_j(u)F_{p_j'}(t,p,p')\right].$$

This yields

$$\frac{d}{dt}L_{\mu'} = \sum_{j=1}^{n}\left[\left(2\mu'N_j(u) + 2\mu(\frac{d}{dt}N_j(u))\right)F_{p_j'}(t,p,p')\right]$$
$$+ \sum_{j=1}^{n}\left[2\mu N_j(u)\frac{d}{dt}F_{p_j'}(t,p,p')\right],$$

and therefore (2.12) can be rewritten as

$$2\mu\sum_{j=1}^{n}\left[N_j(u)\left(F_{p_j}(t,p,p') - \frac{d}{dt}F_{p_j'}(t,p,p')\right)\right] = 0.$$

Summarizing the discussion above, the Euler Lagrange differential equations (2.11) and (2.12) imply the following system of differential equations, where we use the notation $\mu := \mu(t)$, $u := u(t)$, $d_i := F_{p_i}(t, p, p') - \frac{d}{dt} F_{p_i'}(t, p, p')$, $i = 1, \ldots, n$, and $c := \lambda(t)$ for an arbitrary but fixed value of $t \in [0, 1]$:

$$\left(\mu^2 \frac{\partial N_1(u)}{\partial u_1} + 1\right) d_1 + \mu^2 \frac{\partial N_2(u)}{\partial u_1} d_2 + \cdots + \mu^2 \frac{\partial N_n(u)}{\partial u_1} d_n - G_{u_1}(u)c = 0,$$

$$\mu^2 \frac{\partial N_1(u)}{\partial u_2} d_1 + \left(\mu^2 \frac{\partial N_2(u)}{\partial u_2} + 1\right) d_2 + \cdots + \mu^2 \frac{\partial N_n(u)}{\partial u_1} d_n - G_{u_2}(u)c = 0,$$

$$\vdots$$

$$\mu^2 \frac{\partial N_1(u)}{\partial u_n} d_1 + \mu^2 \frac{\partial N_2(u)}{\partial u_n} d_2 + \cdots + \left(\mu^2 \frac{\partial N_n(u)}{\partial u_n} + 1\right) d_n - G_{u_n}(u)c = 0,$$

$$2\mu \left[\; N_1(u)d_1 + N_2(u)d_2 + \cdots + N_n(u)d_n \;\right] = 0.$$

This system can be written as

$$
\begin{aligned}
\left(\mu^2 \, \nabla N(u) + I_n\right) \, D - \nabla G(u)c &= \underline{0} \\
2\mu \, N(u) \, D &= 0,
\end{aligned}
\tag{2.15}
$$

where $D := (d_1, \ldots, d_n)^T$ and I_n denotes the $n \times n$ identity matrix. Recall that $N(u) = \nabla G(u)$, which implies that (2.15) is equivalent to

$$
\begin{aligned}
\left(\mu^2 \, H(G(u)) + I_n\right) \, D - \nabla G(u)c &= \underline{0} \\
2\mu \, \nabla G(u) \, D &= 0,
\end{aligned}
$$

where $H(G(u))$ denotes the Hessian of G at the point u.

In the following we will distinguish the two cases that $\mu(t) \neq 0$ and that $\mu(t) = 0$ for an arbitrary but fixed value of $t \in [0, 1]$:

Case 1: $\mu(t) \neq 0$.

In this case the system (2.15) can be interpreted as a system of $(n + 1)$ linear equations of $d_1, \ldots, d_n, c$ given by $A(d_1, \ldots, d_n, c)^T = \underline{0}$.

(a) Without loss of generality we can assume that A is nonsingular; i.e., $\text{rank}(A) = n + 1$. If this is not the case, we will show in part 1(b) how the problem can be transformed so that A becomes nonsingular.

Since A is nonsingular, (2.15) can be satisfied only if $(d_1, \ldots, d_n, c) = \underline{0}$. This implies that $\lambda(t) = 0$ and that $p(t)$ satisfies the Euler Lagrange differential equations

$$F_{p_j}(t, p, p') - \frac{d}{dt} F_{p_j'}(t, p, p') = 0, \quad j = 1, \ldots, n. \tag{2.16}$$

(b) Now assume that A is singular; i.e., $\mathrm{rank}(A) < n + 1$.

This is a degenerate case that can be avoided by using a representation for p that is a slight modification of (2.7).

First note that a coordinate transformation not changing the problem in any essential way can be applied to the given problem such that in the resulting problem, the normal vector $N(u) = \nabla G(u)$ has any predefined direction. (Recall that $\nabla G(u)$ is different from $\underline{0}$ for every $u \in \partial(B)$.)

Using this observation we can conclude that the submatrix $\bar{A} := \mu^2 \nabla N(u) + I_n$ of A is not regular. This implies further on that the matrix $\nabla N(u)$ has the eigenvalue $-1/\mu^2$.

Additionally, a new representation of $p(t)$ that is equivalent to (2.7) can be defined by

$$p(t) = u(t) + \tilde{\mu}^2(t)\tilde{N}(u(t)), \quad u(t) \in \partial(B), \ \tilde{\mu}(t) \in \mathbb{R},$$

$$\text{s.t. } G(u(t)) = 0 \qquad\qquad \forall t \in [0,1],$$

$$\tag{2.17}$$

where $\tilde{N}(u(t)) := \alpha(u(t))\nabla G(u(t)) = \alpha(u(t))N(u(t))$ with a given continuously differentiable function $\alpha : \mathbb{R}^n \to \mathbb{R}$ satisfying $\alpha(v) > 0$ for all $v \in \mathbb{R}^n$.

Moreover, the function α can be chosen such that $\alpha(u(t)) = 1$, implying that $\tilde{\mu}(t) = \mu(t)$ for the given arbitrary, but fixed, value of t. It is easy to verify that using representation (2.17), a system of differential equations similar to (2.15) is obtained, where $N(u)$ is replaced by $\tilde{N}(u)$ and $\nabla N(u)$ is replaced by $\nabla \tilde{N}(u)$ with

$$\frac{\partial \tilde{N}(u)}{\partial u_i} = \frac{\partial \alpha(u)}{\partial u_i} N(u) + \alpha(u)\frac{\partial N(u)}{\partial u_i} \qquad \forall i = 1,\ldots,n.$$

Obviously, the function α can be chosen such that $\nabla \tilde{N}(u)$ does not have the eigenvalue $-1/\tilde{\mu}^2 = -1/\mu^2$. Combining a suitable coordinate transformation of the problem with an appropriate choice of the function α in (2.17) yields a regular system (2.15), and we can return to the case 1(a) above.

Case 2: $\mu(t) = 0$.

If $\mu(t) = 0$, the system (2.15) is satisfied if and only if

$$F_{p_i}(t, p, p') - \frac{d}{dt} F_{p'_i}(t, p, p') + \lambda(t) G_{u_i}(u) = 0, \quad i = 1, \ldots, n.$$

Since $p(t) = u(t)$ holds in this case (compare (2.7)), this is equivalent to

$$F_{p_i}(t, p, p') - \frac{d}{dt} F_{p'_i}(t, p, p') + \lambda(t) G_{p_i}(p) = 0, \quad i = 1, \ldots, n,$$
$$(2.18)$$

which is the Euler Lagrange differential equation for a curve $p(t)$ on the boundary of B, i.e., for a curve $p(t)$ that satisfies the side constraint $G(p(t)) = 0$.

Summarizing the discussion above we obtain the following result:

Theorem 2.1. *A d-shortest permitted path in $\mathbb{R}^n \setminus \text{int}(\mathcal{B})$ connecting two feasible points X and Y and given by the parameterization p consists of pieces of arcs where*

1. p satisfies the Euler Lagrange differential equations (2.16), or

2. p is a geodesic curve on the boundary of a barrier that satisfies (2.18).

Theorem 2.1 gives a necessary optimality condition for d-shortest permitted paths in $\mathbb{R}^n$. Note that in the unconstrained case where no barriers are given in $\mathbb{R}^n$, and under some convexity assumptions, sufficient optimality conditions for d-shortest paths could also be proven (see, for example, Clarke, 1983; Elsgolc, 1962).

In the case of Euclidean distances, Theorem 2.1 implies the following intuitive optimality condition for l_2-shortest permitted paths:

Corollary 2.2. *An l_2-shortest permitted path in $\mathbb{R}^n \setminus \text{int}(\mathcal{B})$ connecting two feasible points X and Y and given by the parameterization p consists of*

1. straight line segments in $\mathbb{R}^n$, and

2. pieces of arcs where p is a geodesic curve on the boundary of a barrier.

As discussed earlier in this section, Theorem 2.1 and Corollary 2.2 can be easily generalized to the case that a finite set of pairwise disjoint, compact, and convex barriers is given in $\mathbb{R}^n$.

A generalization of the proof of Theorem 2.1 to polyhedral barrier sets is yet more difficult. Recall that the reformulation (2.7) relies on the fact that the boundary of the given barrier B has a representation $\partial(B) =$

$\{X \in \mathbb{R}^n : G(X) = 0\}$ with a twice continuously differentiable function $G : \mathbb{R}^n \to \mathbb{R}$. Since every compact and convex polyhedral barrier set can be approximated arbitrarily well by a smooth barrier set satisfying the above requirements, we can expect that Theorem 2.1 also transfers to this case. A constructive proof of this conjecture will be given in Section 2.3; see Lemma 2.4 on page 31. An example for an l_2-shortest permitted path in $\mathbb{R}^3$ satisfying the necessary optimality condition of Corollary 2.2 is given in Figure 2.5.

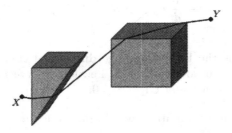

Figure 2.5. An l_2-shortest permitted X-Y path in an environment with two polyhedral barriers in $\mathbb{R}^3$.

2.3   Piecewise Linear Paths for Polyhedral Barriers

In this section we concentrate on the case that a finite set of pairwise disjoint, closed, polyhedral, but not necessarily convex barriers $\{B_1, \ldots, B_N\}$ is given in the n-dimensional real space $\mathbb{R}^n$. Moreover, the number of faces of the barriers in $\mathcal{B} = \bigcup_{i=1}^{N} B_i$ is assumed to be finite to avoid degenerate cases as illustrated in Figure 2.1 on page 17. In the 2-dimensional case, which will be given special attention, the finite set of extreme points of $\mathcal{B}$ is denoted by $\mathcal{P}(\mathcal{B})$.

Observe that using the same arguments as in the previous section we could restrict our analysis to the case that only one barrier set is given. However, since the general case of a finite number of barrier sets does not complicate the following discussion as it did in the previous section, we will not use this simplification here.

Parts of the results discussed in this and in the following sections of this chapter were also published in Dearing et al. (2002), Hamacher and Klamroth (2000), and Klamroth (2001b).

In the 2-dimensional case where distances are measured by an l_p metric with $1 \leq p \leq \infty$, Viegas and Hansen (1985) showed that for every pair of points $X, Y \in \mathcal{F}$, $X \neq Y$, there always exists an l_p-shortest permitted path connecting X and Y that is a piecewise linear path with breaking points only in extreme points of a barrier. The following result shows that this property also holds for any other metric d that is induced by a norm.

Lemma 2.3. *Let $\{B_1, \ldots, B_N\}$ be a finite set of pairwise disjoint, closed, polyhedral barrier sets in $\mathbb{R}^2$ with a finite set of extreme points $\mathcal{P}(\mathcal{B})$. Let d be a metric induced by a norm and let $X, Y \in \mathcal{F}$. Then there exists a d-shortest permitted path SP connecting X and Y with the following property:*

Barrier Touching Property (BTP):
SP is a piecewise linear path with breaking points only in extreme points of barriers.

Proof. Let $X, Y \in \mathcal{F}$ and let SP be any d-shortest permitted path connecting X and Y in $\mathcal{F}$ that does not satisfy the barrier touching property. Note that since the set of barriers and also the set of extreme points $\mathcal{P}(\mathcal{B})$ of the barriers are finite, SP can be subdivided by a finite set of points so that two consecutive points on SP are l_2-visible. Lemma 2.2 therefore implies that the straight line segment connecting two consecutive points on SP is a d-shortest permitted path connecting these two points. We can therefore construct a piecewise linear path SP' with a finite set of breaking points with length less than or equal to that of SP. Now a d-shortest permitted path SP'' with the barrier touching property can be constructed from SP', a construction similar to that given in Viegas and Hansen (1985) for l_p-distances: Let $[T_{i-1}, T_i]$ and $[T_i, T_{i+1}]$ be two consecutive straight line segments of SP'. First assume that T_{i-1} and T_{i+1} are l_2-visible. Then the triangle inequality implies that the two segments $[T_{i-1}, T_i]$ and $[T_i, T_{i+1}]$ can be replaced by one straight line segment $[T_{i-1}, T_{i+1}]$ without increasing the length of SP'. Otherwise, using again the triangle inequality, the breaking point T_i can be moved along $[T_{i-1}, T_i]$ or along $[T_i, T_{i+1}]$ towards T_{i-1} or T_{i+1}, respectively, without increasing the length of SP', until one of these line segments becomes tangent to a barrier.

During the iteration of both operations every extreme point of a barrier that lies on SP' is interpreted as a breaking point T_i even if the straight line segment $[T_{i-1}, T_{i+1}]$ is part of SP'. Thus the iteration of both operations yields a path SP'' with the desired property after a finite number of steps, since every breaking point of SP' that is not yet an extreme point of a barrier can be moved towards X, Y, or an extreme point of a barrier. $\square$

Note that the barrier touching property is in general not satisfied for problems in $\mathbb{R}^n$ with $n \geq 3$. For an example where no l_2-shortest permitted X-Y path in $\mathbb{R}^3$ with breaking points only in extreme points of barriers exists, see Figure 2.5 on page 30. Nevertheless, Lemma 2.3 can be generalized to the n-dimensional case as will be shown in Lemma 2.4 below. An alternative proof of this result for the special case of the Euclidean metric l_2 and for convex barrier sets can be found in De Berg et al. (1995).

Lemma 2.4. *Let $\{B_1, \ldots, B_N\}$ be a finite set of pairwise disjoint, closed, polyhedral sets in $\mathbb{R}^n$, $n \geq 3$, with a finite number of faces. Let d be a metric induced by a norm and let $X, Y \in \mathcal{F}$. Then there exists a d-shortest permitted path SP connecting X and Y with the following property:*

Generalized Barrier Touching Property:
SP is a piecewise linear path with breaking points only in faces of barriers of dimension $n - 2$ or less.

Proof. Let $X, Y \in \mathcal{F}$ and let SP be any d-shortest permitted path connecting X and Y in $\mathcal{F}$ that does not satisfy the generalized barrier touching property. Following the same arguments as in the proof of Lemma 2.3, a piecewise linear permitted X-Y path SP' with $l(SP') \leq l(SP)$ can be constructed by triangulating the feasible region $\mathcal{F}$ such that a finite set of simplices is obtained. According to Lemma 2.2, curved portions of SP in these simplices can be replaced by line segments without increasing the length of SP.

Consider two consecutive line segments $[T_{i-1}, T_i]$ and $[T_i, T_{i+1}]$ on SP'. Observe that $[T_{i-1}, T_i] \cap [T_i, T_{i+1}] = \{T_i\}$, since otherwise SP' would not be a d-shortest permitted X-Y path. Therefore, these line segments uniquely define a plane H in $\mathbb{R}^n$ containing both segments $[T_{i-1}, T_i]$ and $[T_i, T_{i+1}]$. The intersection of H with the barriers in $\mathcal{B}$ results in 2-dimensional polyhedral barrier sets in H with a finite number of extreme points, since the barriers in $\mathcal{B}$ have a finite number of faces. Now Lemma 2.3 can be applied to the 2-dimensional subproblem of finding a d-shortest permitted path connecting the points T_{i-1} and T_{i+1} in the plane H. Therefore, the path $[T_{i-1}, T_i] \cup [T_i, T_{i+1}]$ can be replaced by a piecewise linear T_{i-1}-T_{i+1} path in H satisfying the barrier touching property (BTP) and having length less than or equal to that of the original T_{i-1}-T_{i+1} path in SP'. Since an extreme point of the polyhedral intersection set of a barrier $B \in \mathcal{B}$ with the plane H is a zero-dimensional face of $B \cap H$, this point corresponds to a point on a face of B in $\mathbb{R}^n$ that has dimension $n - 2$ or less. We can conclude that the constructed T_{i-1}-T_{i+1} path satisfies the generalized barrier touching property.

Iterating this procedure a finite number of times thus allows the construction of a d-shortest permitted X-Y path with breaking points only in faces of barriers of dimension $n - 2$ or less. $\square$

2.4   Shortest Paths in the Plane with Polyhedral Barriers

Since many applications in location planning deal with planar environments, planar models are of prime importance in this context. Moreover, in a major part of applications the barrier regions can be approximated by polyhedral sets without significantly influencing the accuracy of the model.

As in Section 2.3, we focus on pairwise disjoint, closed, polyhedral, but not necessarily convex, barriers $\{B_1, \ldots, B_N\}$ in the plane $\mathbb{R}^2$. The union of the barrier regions is denoted by $\mathcal{B} = \bigcup_{i=1}^{N} B_i$, and the finite sets of extreme points and facets of $\mathcal{B}$ are denoted by $\mathcal{P}(\mathcal{B})$ and $\mathcal{F}(\mathcal{B})$, respectively.

An immediate consequence of Lemma 2.3 is that the barrier distance $d_\mathcal{B}(X, Y)$ for an arbitrary pair of points $X, Y \in \mathcal{F}$ can be calculated with respect to an *intermediate point* $I_{X,Y}$:

Definition 2.5. *Let $\{B_1, \ldots, B_N\}$ be a finite set of pairwise disjoint, closed, polyhedral sets in $\mathbb{R}^2$ with a finite set of extreme points, and let $X, Y \in \mathcal{F}$, $X \neq Y$.*

An intermediate point $I_{X,Y}$ *is a point different from Y that is a breaking point or an extreme point of a barrier that lies on a d-shortest permitted X-Y path with the barrier touching property and that is d-visible from Y.*

If X and Y are d-visible, the intermediate point $I_{X,Y}$ can be chosen as X. If $X = Y$, then $I_{X,Y} := X$.

Note that an intermediate point $I_{X,Y}$ is not necessarily unique. Consider, for example, the case that X and Y are d-visible. Then an intermediate point $I_{X,Y}$ may be the point X, but it may also be a breaking point or an extreme point of a barrier on a d-shortest permitted X-Y path, which is not necessarily a straight line segment (cf. Definition 2.2 on page 17). Moreover, as a result of Lemmas 2.2 and 2.3, an intermediate point $I_{X,Y}$ can always be chosen such that it is not only d-visible from Y but also l_2-visible from Y, as can be seen in Figure 2.6.

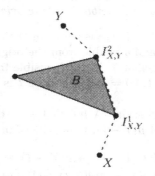

Figure 2.6. Two intermediate points on an l_1-shortest permitted X-Y path with the barrier touching property. Note that $I_{X,Y}^1$ is only l_1-visible from Y, whereas $I_{X,Y}^2$ is l_1-visible and l_2-visible from Y.

We say that an intermediate point $I_{X,Y}$ *is assigned to X* if a d-shortest permitted X-Y path with the barrier touching property passes through $I_{X,Y}$.

Corollary 2.3. *Let d be a metric induced by a norm and let $X, Y \in \mathcal{F}$. Furthermore, let SP be a d-shortest permitted X-Y path with the barrier touching property and let the point $I_{X,Y}$ be an intermediate point according to Definition 2.5. Then*

$$d_\mathcal{B}(X, Y) = d_\mathcal{B}(X, I_{X,Y}) + d(I_{X,Y}, Y).$$

To obtain a similar representation of permitted X-Y paths for higher-dimensional problems, auxiliary vertices can be placed on the boundary segments (in $\mathbb{R}^3$) and on the faces of dimension $n - 2$ and less (in $\mathbb{R}^n$, $n > 3$) of the polyhedral barrier sets having a certain maximum distance from each other. Even though this approach allows a good approximation of d-shortest permitted X-Y paths in $\mathbb{R}^n$ by piecewise linear paths with breaking points only in the specified vertices of the barriers (see, for example, Choi et al. (1995) and Lozano-Perez and Wesley (1979) for the three-dimensional case), the complexity of the resulting problem strongly depends on the number of added vertices and therefore on the desired accuracy of the approximation.

Another consequence of Lemmas 2.2 and 2.3 is that the boundary $\partial(\text{shadow}_d(X))$ of the closure of $\text{shadow}_d(X)$ (cf. Definition 2.4) is piecewise linear for any distance function d induced by a norm. The following technical lemma will be useful for the proof of this statement. For simplicity we assume without loss of generality that $X = \underline{0}$ is the origin.

Lemma 2.5. *Let d be a metric induced by a norm. Furthermore, let $Y \in \mathcal{F}$ be a point that is d-visible from the origin and let $I := I_{\underline{0},Y}$ be an intermediate point on a d-shortest permitted $\underline{0}$-Y path with the barrier touching property. Let $Z = I + \lambda(Y - I)$, $\lambda \geq 0$, be a point in $\mathcal{F}$ such that Z is l_2-visible from I. Then Z is d-visible from the origin.*

Proof. First assume that Y is l_2-visible from the origin. Then $I = \underline{0}$, and thus Z is also l_2-visible and d-visible from the origin.

Now consider the case that Y is not l_2-visible from the origin and thus $I \neq \underline{0}$. Then $d(\underline{0}, I) + d(I, Y) = d(\underline{0}, Y)$, since I is a point on a d-shortest permitted $\underline{0}$-Y path.

Assume that there exist scalars $\lambda, \mu \in [0, 1]$ such that $d(\underline{0}, Z) < \lambda d(\underline{0}, I) + \mu d(I, Y)$, where $Z = \lambda I + \mu(Y - I)$. Using the triangle inequality we obtain

$$
\begin{aligned}
d(\underline{0}, Y) &= d(\underline{0}, \lambda I + (1 - \lambda)I + \mu(Y - I) + (1 - \mu)(Y - I)) \\
&\leq d(\underline{0}, Z) + (1 - \lambda)d(\underline{0}, I) + (1 - \mu)d(I, Y) \\
&< \lambda d(\underline{0}, I) + \mu d(I, Y) + (1 - \lambda)d(\underline{0}, I) + (1 - \mu)d(I, Y) \\
&= d(\underline{0}, I) + d(I, Y),
\end{aligned}
$$

contradicting the assumption that I is a point on a d-shortest $\underline{0}$-Y path.

Thus $d(\underline{0}, Z) = \lambda d(\underline{0}, I) + \mu d(I, Y)$ for all $\lambda, \mu \in [0, 1]$, which, using $\lambda = 1$, proves the result for all $\mu \in [0, 1]$.

The remaining case is that $\lambda = 1$ but $\mu > 1$, i.e., that $Z = I + \mu(Y - I)$, $\mu > 1$. Assume that there exists a scalar $\mu > 1$ such that $d(\underline{0}, Z) < d(\underline{0}, I) + \mu d(I, Y)$. It follows that

$$\mu d\left(\underline{0}, \frac{1}{\mu}I + (Y - I)\right) < d(\underline{0}, I) + \mu d(I, Y),$$

which is equivalent to

$$d\left(\underline{0}, \frac{1}{\mu}I + (Y - I)\right) < d(\underline{0}, \frac{1}{\mu}I) + d(I, Y).$$

Using the inequalities derived for the previously discussed case, this completes the proof. $\qquad\square$

Lemma 2.5 implies that the boundary of the closure of $\text{shadow}_d(X)$ is piecewise linear for all points $X \in \mathcal{F}$:

Lemma 2.6. *Let d be a metric induced by a norm and let $X \in \mathcal{F}$ be a feasible point. Then $\partial(\text{shadow}_d(X))$ is piecewise linear.*

Proof. Those parts of $\partial(\text{shadow}_d(X))$ that are part of the boundary of a barrier region are obviously piecewise linear, since all barrier sets are assumed to be polyhedral sets. For all other parts of $\partial(\text{shadow}_d(X))$, consider a point Y on $\partial(\text{shadow}_d(X))$ and let $I_{X,Y}$ be an intermediate point on a d-shortest permitted X-Y path with the barrier touching property. Note that in this case Y is d-visible from X. If for all possible choices of Y, all the points Z on the line segment starting at $I_{X,Y}$ passing through Y and ending as soon as the interior of a barrier is intersected are d-visible from X and thus not in $\text{shadow}_d(X)$, then $\partial(\text{shadow}_d(X))$ is piecewise linear. Since we can assume without loss of generality that $X = \underline{0}$ is the origin, Lemma 2.5 implies the result. $\qquad\square$

Lemma 2.6 implies that $\text{shadow}_d(X)$ has an analytic representation for all $X \in \mathcal{F}$. Note that $\text{shadow}_d(X)$ is nevertheless not necessarily convex, as can be seen in the example in Figure 2.2 on page 19.

2.5 Shortest Paths and Polyhedral Gauges in the Plane

In addition to the general assumptions made in Section 2.1, we assume in the following that a polyhedral gauge γ is the given metric in the plane, defined by its unit ball P according to Definition 1.6. In particular, we assume that P is a compact convex polytope with extreme points $\text{ext}(P) = \{v^1, \ldots, v^\delta\}$, given in clockwise order, and that $v^{\delta+1} = v^1$. The corresponding fundamental directions are denoted by $d^1, \ldots, d^\delta$.

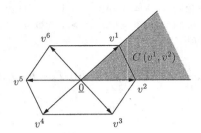

Figure 2.7. A (symmetric) polyhedral gauge with six fundamental vectors.

Recall that Lemma 1.5 on page 9 implies that only the two fundamental vectors v^i and v^{i+1} need to be used to determine the value of $\gamma(X)$ for a point X in the cone $C\left(v^i, v^{i+1}\right)$.

We consider the situation where only one convex, compact barrier $\mathcal{B} = B$ is given in $\mathbb{R}^2$ that can not be trespassed. In this situation we consider a point $X \in C\left(v^i, v^{i+1}\right)$, see Figure 2.7, and distinguish three cases in which $B \cap C\left(v^i, v^{i+1}\right) \neq \emptyset$: see Figure 2.8:

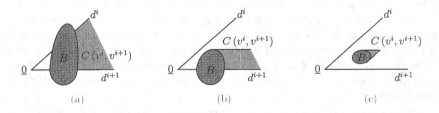

Figure 2.8. Three cases in which a barrier B changes the distance between the origin and a point $X \in \mathcal{F}$. The shaded area is that part of $C\left(v^i, v^{i+1}\right)$ that lies in the γ-shadow of $\underline{0}$; i.e., $\gamma_B(\underline{0}, X) > \gamma(\underline{0}, X)$.

Case (a)   The fundamental directions d^i and d^{i+1} both contain points of B.

Case (b)   Only one of the fundamental directions, say d^{i+1}, contains a point of B.

Case (c)   Neither d^i nor d^{i+1} contains points of B.

In all cases B partitions $C\left(v^i, v^{i+1}\right) \cap \mathcal{F}$ into two parts: one part in which there exists a permitted path from $\underline{0}$ to X with length $\gamma(\underline{0}, X)$ and one part that is part of shadow$_\gamma(\underline{0})$.

In all three cases the non γ-visible part of $C\left(v^i, v^{i+1}\right)$ has to be determined differently. The non γ-visible part of $C\left(v^i, v^{i+1}\right)$ in Case (a) equals the non l_2-visible part of $C\left(v^i, v^{i+1}\right)$. In Case (b) we have $B \cap d^{i+1} \neq \emptyset$. Then the non γ-visible part of $C\left(v^i, v^{i+1}\right)$ is a subset of the non l_2-visible part of $C\left(v^i, v^{i+1}\right)$. It is the region bounded by $\partial(B)$, d^{i+1} and the "tangent" to B in $C\left(v^i, v^{i+1}\right)$ parallel to d^{i+1}. Analogously, in Case (c) the non γ-visible points are non l_2-visible, and the corresponding subset of $C\left(v^i, v^{i+1}\right)$ is bounded by $\partial(B)$ and two tangents on $\partial(B)$ parallel to d^i and d^{i+1}, respectively. Note that in all cases the set of non γ-visible points is also non l_2-visible.

In the special case of block norms (i.e., symmetric polyhedral gauges) and polyhedral barrier sets, an interesting question that was already addressed in Section 1.3 is that of possible interrelations between block norms with four fundamental vectors (for an example see Figure 1.7 on page 12) and the Manhattan norm l_1. As was shown in Lemma 1.6 on page 13, these distance measures can be viewed as mathematically equivalent in the unconstrained case. In the following we will generalize Lemma 1.6 so that it can also be applied to problems involving barriers.

For this purpose let for the remainder of this section $\mathcal{B} = \bigcup_{i=1}^{N} B_i$ be the union of a finite set of pairwise disjoint, closed, and polyhedral barrier sets in $\mathbb{R}^2$. By $v^1, \ldots, v^4$ we denote the fundamental vectors of the l_1 norm (see Figure 1.5 on page 11), and we assume that γ is a block norm in $\mathbb{R}^2$ with the four fundamental vectors (in clockwise order)

$$u^1 = \left(u_1^1, u_2^1\right)^T, \quad u^2 = \left(u_1^2, u_2^2\right)^T, \quad u^3 = \left(u_1^3, u_2^3\right)^T, \quad u^4 = \left(u_1^4, u_2^4\right)^T,$$

such that $u^3 = -u^1$ and $u^4 = -u^2$.

Furthermore, a linear transformation T with $T\left(u^1\right) = Tu^1 = v^1$ and $T\left(u^2\right) = Tu^2 = v^2$ is defined as in (1.26). By $T(\mathcal{B})$ and $T(\mathcal{F})$ we denote the image of $\mathcal{B}$ and of the feasible region $\mathcal{F}$ under the linear transformation T, respectively. Note that T is linear and nonsingular, since the fundamental vectors u^1 and u^2 of a block norm are linearly independent. Thus the set $T(\mathcal{B})$ in the image space is again the union of a finite set of pairwise disjoint, closed, polyhedral barriers. In particular, T is a linear bijective mapping that defines a one-to-one correspondence between the extreme points of $\mathcal{B}$ in the original space and the extreme points of $T(\mathcal{B})$ in the image space. The corresponding inverse mapping is denoted by T^{-1}.

Using Lemmas 2.2 and 2.3 we can derive the following relation between the barrier distances $\gamma_{\mathcal{B}}$ and $l_{1, T(\mathcal{B})}$:

Lemma 2.7. *Let γ be a block norm with four fundamental vectors and let the linear transformation T be defined as in (1.26). Then for all $X, Y \in \mathcal{F}$,*

$$\gamma_{\mathcal{B}}(X, Y) = l_{1, T(\mathcal{B})}(T(X), T(Y)).$$

Proof. Let $X, Y \in \mathcal{F}$ be two feasible points and let SP be a γ-shortest permitted path connecting X and Y with the barrier touching property of Lemma 2.3. Thus SP is a piecewise linear path with a finite number of breaking points $I_1, \ldots, I_k \in \mathcal{F}$. Since T is linear and bijective, the image $T(SP)$ of SP is feasible in $T(\mathcal{F})$; i.e., $T(SP) \subseteq T(\mathcal{F})$. Furthermore, line segments in SP are transformed into line segments in $T(SP)$, and thus $T(SP)$ is also a piecewise linear path with breaking points $T(I_i)$, $i = 1, \ldots, k$. Setting $I_0 := X$ and $I_{k+1} := Y$ and using Lemma 1.6, we get

$$\gamma_{\mathcal{B}}(X, Y) = \sum_{i=0}^{k} \gamma(I_i, I_{i+1}) = \sum_{i=0}^{k} l_1(T(I_i), T(I_{i+1}))$$
$$\geq l_{1, T(\mathcal{B})}(T(X), T(Y)).$$

Analogously, let SP' be an l_1-shortest permitted path with the barrier touching property in $T(\mathcal{F})$, connecting $T(X)$ and $T(Y)$. Let $I'_1, \ldots, I'_{k'} \in T(\mathcal{F})$ be the finite set of breaking points on SP' and let $I'_0 := T(X)$ and $I'_{k'+1} := T(Y)$. Then

$$l_{1, T(\mathcal{B})}(T(X), T(Y)) = \sum_{i=0}^{k'} l_1(I'_i, I'_{i+1}) = \sum_{i=0}^{k'} \gamma\left(T^{-1}(I'_i), T^{-1}(I'_{i+1})\right)$$
$$\geq \gamma_{\mathcal{B}}(X, Y),$$

completing the proof. □

Analogously, it can be shown that under the assumption of Lemma 2.7,

$$l_{1, \mathcal{B}}(X, Y) = \gamma_{T^{-1}(\mathcal{B})}\left(T^{-1}(X), T^{-1}(Y)\right).$$

Lemma 2.7 implies that the problem of finding γ-shortest permitted paths with respect to a block norm γ having four fundamental directions can be reduced to the problem of finding l_1-shortest permitted paths in a linearly transformed environment. As will be discussed in the following section, efficient algorithms are available in this case for the computation of shortest permitted paths, particularly in the presence of polyhedral barriers.

3

Location Problems with Barriers:
Basic Concepts and Literature Review

In the previous chapters we have dealt with the question of how best to define a distance between two fixed points in the n-dimensional real space $\mathbb{R}^n$ if constraints for traveling in the form of barriers are given. In this chapter we are free to choose one of these points in space, a *new location*, as long as we do not interfere with the given barriers.

3.1 Location Problems Without Barriers

In times of increasing transportation costs and just-in-time delivery schedules, good locational decisions are needed in many different fields. Decisions in management, economy, production planning, etc., contain many facets that are related to "locating facilities." The location of a new warehouse with respect to a given set of customers or the location of an emergency facility in an expanding neighborhood are only two examples for a wide range of applications. On the other hand, location theory is in its own right an interesting and challenging part of mathematics with an ever increasing set of problems that do not necessarily have a real-world background.

In a general *continuous location model*, a finite set of *existing facilities*, or customers, is given, represented by points in the n-dimensional space $\mathbb{R}^n$. If $n = 2$, a *planar location model* is obtained, a special case of particular importance, since practical applications very often deal with locational decisions in the plane.

The set of existing facilities is denoted by $\mathcal{E}x := \{Ex_1, \ldots, Ex_M\}$ with the index set $\mathcal{M} := \{1, \ldots, M\}$, where $Ex_m \in \mathbb{R}^n$ for all $m \in \mathcal{M}$.

Moreover, a *distance measure* $d : \mathbb{R}^n \times \mathbb{R}^n \to \mathbb{R}$ is given, modeling, for example, the travel cost or the travel time between two points $X, Y \in \mathbb{R}^n$. We will assume in the following that d is a metric induced by a norm $\|\bullet\|_d$ in $\mathbb{R}^n$; that is, d satisfies (1.7), (1.8), and (1.9); see page 5.

Then the *continuous single-facility location problem* is to locate one new facility $X \in \mathbb{R}^n$ such that some function of the distances between the new facility X and the existing facilities in $\mathcal{E}x$ is minimized:

$$\min \quad f(X) = f(d(X, Ex_1), \ldots, d(X, Ex_M))$$
$$\text{s.t.} \quad X \in \mathbb{R}^n. \tag{3.1}$$

We will often assume that f is a convex and nondecreasing function of the distances $d(X, Ex_m)$, $m \in \mathcal{M}$.

A well-known example is the *Weber problem* (also often referred to as the *Fermat Steiner Weber problem*, *median problem*, *minisum problem*, or, in the special case of three existing facilities having equal weights, as the *Fermat Torricelli problem*):

$$\min \quad f(X) = \sum_{m=1}^{M} w_m d(X, Ex_m)$$
$$\text{s.t.} \quad X \in \mathbb{R}^n, \tag{3.2}$$

where the nonnegative weight $w_m = w(Ex_m)$ specifies the demand of the existing facility Ex_m, $m \in \mathcal{M}$. Thus the objective is to minimize the total travel cost between the new facility X and $\mathcal{E}x$, modeling, for example, a warehouse, and a set of customers, respectively.

Weber problems with different distance functions and various extensions and variations have been extensively studied in the literature. For an overview of the Weber problem and related problems, see, for example, Wesolowsky (1993) and Drezner et al. (2002).

If, on the other hand, the location for an emergency facility such as a hospital or a fire station is sought, the *center problem* (*Weber Rawls problem*, *minimax problem*)

$$\min \quad f(X) = \max_{m \in \mathcal{M}} w_m d(X, Ex_m)$$
$$\text{s.t.} \quad X \in \mathbb{R}^n \tag{3.3}$$

can be used as an appropriate model, where, as in (3.2), the weights w_m, $m \in \mathcal{M}$, are assumed to be nonnegative. A survey of the center problem and of related covering location problems is given, for example, in Plastria (2002).

Note that in both cases existing facilities Ex_m having zero demand $w_m = 0$ can be omitted from the model without changing the objective function. Therefore, we will often assume without loss of generality that all the weights w_m, $m \in \mathcal{M}$, are strictly positive.

Both problems (3.2) and (3.3) can be seen as special cases of the more general concept of *ordered Weber problems* as introduced in Puerto and Fernández (2000): For nonnegative weights w_m, $m \in \mathcal{M}$, the values $d_m(X) := w_m d(X, Ex_m)$ are defined and sorted using a permutation π : $\{1, \ldots, M\} \to \{1, \ldots, M\}$ such that

$$d_{\pi(1)}(X) \le d_{\pi(2)}(X) \le \cdots \le d_{\pi(M)}(X).$$

Then an ordered Weber problem is defined with respect to an M-dimensional vector $\lambda = (\lambda_1, \ldots, \lambda_M)$ of nonnegative scalars as

$$
\begin{aligned}
\min \quad & f_\lambda(X) = \sum_{m=1}^{M} \lambda_m d_{\pi(m)}(X) \\
\text{s.t.} \quad & X \in \mathbb{R}^n.
\end{aligned}
\tag{3.4}
$$

Note that the Weber problem (3.2) is obtained if the scalars in λ are chosen as $\lambda_m = 1$ for all $m \in \mathcal{M}$, and the center problem (3.3) is obtained for $\lambda_m = 0$, $m = 1, \ldots, M - 1$, and $\lambda_M = 1$. Moreover, (3.4) is equivalent to a *cent-dian problem* if $\lambda_m = \omega$ for $m = 1, \ldots, M - 1$ with some $w \in (0, 1)$, and $\lambda_M = 1$.

A detailed introduction to continuous location theory can be found in the textbooks of Drezner (1995), Francis et al. (1992), Hamacher (1995), Love et al. (1988), and in the survey of Plastria (1995), among others.

Even though we will mostly concentrate on continuous location models in the following chapters, we will sometimes also refer to *network location problems*. In such problems, a simple, finite graph G with vertex set $V = V(G)$ and edge set $E = E(G)$ is given, and demands and travel between existing facilities and new facilities are assumed to occur only on the network G. In particular, a (nonnegative) cost $c(e) = d(v_i, v_j)$ is associated with every edge $e = \{v_i, v_j\} \in E$ connecting two nodes $v_i, v_j \in V$, representing, for example, the distance between the nodes v_i and v_j. The *network distance* $d_G(v_i, v_j)$ between two arbitrary nodes of G is defined as the length of a shortest network path in G. Similarly, the network distance between a point on an edge of G and an arbitrary point on G can be determined by considering the corresponding portions of edges contained in a shortest network path between the two points.

For a given finite set of existing facilities in the vertices of G, a *network location problem* is to find a point X either in a node or on an edge of G such that some objective function $f_G(X) = f_G(d_G(X, Ex_1), \ldots, d_G(X, Ex_M))$ of the network distances between X and the existing facilities $Ex_1, \ldots,$ $Ex_M \in V(G)$ is minimized. Analogously, the *node network location problem* is to find a node in $V(G)$ such that $f_G(v)$ is minimized over all nodes $v \in V(G)$. Well-known examples for possible choices of the objective function f_G are again the Weber objective function (cf. (3.2)) and the center objective function (cf. (3.3)), but also ordered Weber problems can be formulated (see Nickel and Puerto, 1999; Rodriguez-Chia et al., 2000).

An overview of different types of network location problems and solution approaches can be found, for example, in the textbooks of Daskin (1995), Handler and Mirchandani (1979), and Mirchandani and Francis (1990).

To simplify further notation we will use the classification $Pos1/Pos2/Pos3/Pos4/Pos5$ of location problems as introduced in Hamacher (1995) and Hamacher and Nickel (1998). In this classification scheme, $Pos1$ gives the number (and shape) of new facilities sought. In the case that we want to find $N \geq 1$ new points $X_1, \ldots, X_N \in \mathbb{R}^n$, this position contains the positive integer N. $Pos2$ denotes the decision space of the location problem, i.e., $\mathbb{R}^n$ for n-dimensional continuous problems, or P (or $\mathbb{R}^2$) for planar problems. Moreover, we use the symbol G in this position to identify network location problems. $Pos3$ is reserved for special features of a given location problem. This position can be used, for example, to emphasize the fact that forbidden regions are given in $\mathbb{R}^n$ where the location of new facilities is prohibited, denoted by $\mathcal{R}$, or that barriers additionaly restrict traveling in $\mathbb{R}^n$, denoted by $\mathcal{B}$. In continuous models, $Pos4$ contains the information about the distance function, for example l_2 in the case of Euclidean distances, l_1 in the case of the Manhattan metric, or $d_\mathcal{B}$ in the case that a barrier distance function induced by a metric d is given. For network location problems, this position can be used to specify whether distances are measured only between nodes of the network $(d_G(V, V))$, or whether a distance measure is defined everywhere on the graph $(d_G(V, E))$. $Pos5$ indicates the objective function. If, for example, a Weber problem (3.2) is considered, the symbol $\sum$ is used in this position. Center problems (3.3) are identified by the symbol max, ordered Weber problems (3.4) by $\sum_{ord}$, and a general convex objective function (3.1) is referred to by f convex. If no special assumptions are made in any of the positions, this is indicated by a bullet, "•".

Summarizing the discussion above, problems (3.1), (3.2), (3.3), and (3.4) are classified as $1/\mathbb{R}^n/\bullet/d/f$ convex, $1/\mathbb{R}^n/\bullet/d/\sum$, $1/\mathbb{R}^n/\bullet/d/$ max, and $1/\mathbb{R}^n/\lambda/d/\sum_{ord}$, respectively. Similarly, the network location problem and the node network location problem are classified as $1/G/\bullet/d_G(V, E)/f$, and $1/G/\bullet/d_G(V, V)/f$, respectively.

3.2 Introducing Barriers to Location Modeling

The development of realistic location models is a crucial phase in every locational decision process. Especially in the case of planar location models we deal with a geometric representation of the problem, and the geographical reality has to be incorporated into this representation. In almost every real-life situation we have to deal with restrictions and constraints of various types. In the context of location modeling, such restrictions are, for example, regions in which the placement of a new facility is forbidden, but transportation is still possible. These regions, often referred to as *forbidden regions*, can be used to model, for example, state parks or other protected areas, or regions where the geographic characteristics such as the slope forbid the construction of the desired facility. For an overview of location problems with forbidden regions, see, for example, Hamacher and Nickel (1995) and Nickel (1995).

Likewise, there often exist regions where not only the location of a new facility is forbidden, but also where traveling is possible only at a higher cost, such as big lakes that can be crossed only using a ferry boat. Location problems with *congested regions* allowing for different travel speeds or travel costs have been discussed, for example, in Butt (1994); Butt and Cavalier (1997).

Moreover, in many areas traveling is completely forbidden or even impossible. To give only some examples of possible *barrier regions*, consider military areas, mountain ranges, lakes, big rivers, highways, or, on a smaller scale, big machines and conveyor belts in an industrial plant. Applications of barrier regions mentioned in the literature involve, for example, circuit board design (LaPaugh, 1980), facility layout (Francis et al., 1992), pipe network design for ships (Wangdahl et al., 1974), and location and routing of robots (Lozano-Perez and Wesley, 1979). Many other applications can be found in Katz and Cooper (1981), Larson and Li (1981), Larson and Sadiq (1983), and Plastria (1995), among others.

Let $\{B_1, \ldots, B_N\}$ be a finite set of closed and pairwise disjoint barrier sets in $\mathbb{R}^n$, and let $\mathcal{B} = \bigcup_{i=1}^{N} B_i$. The interior of these barrier regions is forbidden for the placement of a new facility, and additionally, traveling through int($\mathcal{B}$) is prohibited. Thus the feasible region $\mathcal{F} \subseteq \mathbb{R}^n$ for new locations and for traveling is given by

$$\mathcal{F} = \mathbb{R}^n \setminus \text{int}(\mathcal{B}).$$

To avoid trivial, infeasible cases we assume, as in the previous chapter, that $\mathcal{F}$ is a connected subset of $\mathbb{R}^n$.

An exception to this definition of $\mathcal{B}$ and $\mathcal{F}$ will be made only in Chapter 7, where *line barriers* are discussed, which have an empty interior. In this case trespassing will be forbidden also at noninterior points of a barrier.

As in the case of unconstrained continuous location problems, a finite set of existing facilities $\mathcal{E}x = \{Ex_m \in \mathbb{R}^n : m \in \mathcal{M}\}$, $\mathcal{M} = \{1, \ldots, M\}$ is given, where we additionally assume that $Ex_m \in \mathcal{F}$ for all $m \in \mathcal{M}$; i.e., all existing facilities are located in the feasible region $\mathcal{F}$. Furthermore, we assume that a metric d induced by a norm $\| \bullet \|_d$ is given by $d(X, Y) = \|Y - X\|_d$, $X, Y \in \mathbb{R}^n$.

The major difference to unconstrained location problems is embodied in the definition of the barrier distance $d_\mathcal{B}$ in Definition 2.2 on page 17. Recall that for d_B the triangle inequality (2.4) is satisfied (see Lemma 2.1 on page 18), but that d_B is in general not positively homogeneous.

Using these general assumptions, the continuous single-facility location problem (3.1) can be restated: While the unconstrained location problem $1/\mathbb{R}^n/\bullet/d/f$ is to find a new facility $X \in \mathbb{R}^n$ minimizing $f(X) = f(d(X, Ex_1), \ldots, d(X, Ex_M))$, the corresponding location problem with barriers $1/\mathbb{R}^n/\mathcal{B}/d_B/f$ can be formulated as

$$
\begin{aligned}
\min \quad & f_B(X) = f(d_B(X, Ex_1), \ldots, d_B(X, Ex_M)) \\
\text{s.t.} \quad & X \in \mathcal{F}.
\end{aligned}
\tag{3.5}
$$

where $f_B(X) = f(d_B(X, Ex_1), \ldots, d_B(X, Ex_M))$ is a function of the distances between a new facility X and the existing facilities in $\mathcal{E}x$. As already indicated in Section 3.1, we will usually assume that f is a convex and nondecreasing function of the distances $d_B(X, Ex_m)$, $m \in \mathcal{M}$.

Special cases of (3.5) include the *Weber problem with barriers* $1/\mathbb{R}^n/\mathcal{B}/d_B/\sum$ given by

$$
\begin{aligned}
\min \quad & f_B(X) = \sum_{m=1}^{M} w_m d_B(X, Ex_m) \\
\text{s.t.} \quad & X \in \mathcal{F}.
\end{aligned}
\tag{3.6}
$$

and the *center problem with barriers* $1/\mathbb{R}^n/\mathcal{B}/d_B/\max$, which can be stated as

$$
\begin{aligned}
\min \quad & f_B(X) = \max_{m \in \mathcal{M}} w_m d_B(X, Ex_m) \\
\text{s.t.} \quad & X \in \mathcal{F}.
\end{aligned}
\tag{3.7}
$$

where, as in (3.2) and (3.3), the weights w_m, $m \in \mathcal{M}$, are assumed to be nonnegative.

The increasing interest in location models incorporating restrictions and barrier regions is reflected in the recent literature. Most of the research so far focused on planar Weber problems with barriers $1/P/\mathcal{B}/d_B/\sum$, considering specific distance functions, barrier shapes, and other modifications of this problem. Problems with other properties and with more general objective functions, including, for example, the center problem (3.7), have only found little attention so far.

Barrier regions were first introduced to location modeling by Katz and Cooper (1981). In this paper the authors consider a problem of the type $1/P/(\mathcal{B} = 1\,circle)/l_{2,B}/\sum$, i.e., a Weber problem in the plane $\mathbb{R}^2$ with the Euclidean metric and with one circular barrier. A heuristic algorithm is suggested that is based on the sequential unconstrained minimization technique (SUMT) for nonlinear programming problems (see Armacost and Mylander, 1973; Fiacco and McCormic, 1968). These results were generalized to the case that several circular barriers are given in the plane, and to the case that one polyhedral barrier set is given in the plane and distances are measured by any l_p metric, $1 < p < \infty$; see Katz and Cooper (1979a,b).

Likewise, for the Weber problem in the plane Aneja and Parlar (1994) and, more recently, Butt (1994) and Butt and Cavalier (1996) developed heuristics for the case that a finite set of polyhedral barriers is given and an l_p metric is used as distance measure. In Aneja and Parlar (1994) problems with polyhedral forbidden regions $1/P/(\mathcal{R} = N\,polyhedra)/l_p/\sum$ and with polyhedral barriers $1/P/(\mathcal{B} = N\,polyhedra)/l_{p,\mathcal{B}}/\sum$ are discussed, where $1 < p \leq 2$. In the barrier case, the visibility graph of the problem is constructed (cf. Section 3.3) and used to evaluate the objective function value at possible solution points. An approximation of the optimal solution is then determined by applying simulated annealing (see, for example, Van Laarhoven and Aarts, 1987). Butt and Cavalier (1996) considered the special case $1/P/(\mathcal{B} = N\,convex\,polyhedra)/l_{2,\mathcal{B}}/\sum$, using the Euclidean metric. They also make use of the visibility graph of the problem and suggest a heuristic optimization scheme, iteratively generating optimal solutions of related unconstrained problems while improving the current solution in each iteration.

Klamroth (2001a) considered the Weber problem for the case that the barrier is a line with a finite number of passages and distances are measured by any metric induced by a norm. A reduction of the nonconvex original problem to a polynomial number of unconstrained Weber problems is given. This approach was extended to the multiple criteria case in Klamroth and Wiecek (2002).

In the special case of the Manhattan metric l_1, discretization results were proven by Larson and Sadiq (1983) for the Weber problem with arbitrarily shaped barriers, $1/P/\mathcal{B}/l_{1,\mathcal{B}}/\sum$. Using a grid tessellation of the plane, an easily determined finite dominating set containing at least one optimal solution to the problem is identified. An implementation of the resulting algorithm in the framework of a geographic information system is presented in Ding et al. (1994). A similar discretization was derived for a more general class of distance functions in Hamacher and Klamroth (2000), namely the classes of block norms and polyhedral gauges. The computational efficiency of these methods was significantly improved by Segars (2000) and Dearing and Segars Jr. (2000a,b), who showed that the consideration of a reduced dominating set is sufficient to solve the problem.

The results of Larson and Sadiq (1983) were also generalized by Batta et al. (1989), who additionally included forbidden regions in the model with the Manhattan metric, and by Savaş et al. (2001) and Wang et al. (2002), who located finite-size facilities acting as barriers themselves. Moreover, Batta et al. (1989) discussed stochastic queue Weber problems as introduced in Berman et al. (1985) and Chiu et al. (1985), adding barriers to the model, and developed connections between network location problems and planar location problems with the Manhattan metric.

For the special case that n pairwise disjoint axis-aligned rectangles are given as barriers, Kusakari and Nishizeki (1997) presented an output-sensitive $O((k + n) \log n)$-time algorithm for the Weber problem with the Manhattan metric, assuming that the number of existing facilities m is a small constant. This result was improved to $O(n \log n + k)$ by Choi et al. (1998), who also gave an $O\left(n^2 m \log^2 m\right)$-time algorithm, based on parametric search, for the center problem. Ben-Moshe et al. (2001) recently improved this last result for the unweighted center problem by giving an $O(nm \log(n+m))$-time algorithm.

Fekete et al. (2000) introduce Weber problems with continuous demand over some given polyhedral set, possibly with holes acting as barriers to travel, and the Manhattan metric. While a polynomial-time algorithm is developed for the case that only one new facility is sought, NP-hardness is proven for the case of multifacility Weber problems if the number of new facilities is part of the input data. Lower and upper bounds as well as the relative accuracy of solutions for multifacility Weber problems with the Manhattan metric, with and without barriers, are also discussed in Batta and Leifer (1988).

Dearing et al. (2002) were, to the best of our knowledge, the first authors who concentrated on center problems with barriers in a more general setting. They derive a dominating set result for polyhedral barriers and the case that distances are measured by the Manhattan metric, and discuss the algorithmic as well as the geometric consequences of this result.

Klamroth (2001b) developed a reduction result that implies a solution strategy for location problems with polyhedral barriers in a very general setting. The objective functions considered are convex functions of distances between a set of existing facilities and one new facility. This definition includes, for example, the Weber and center objective functions as well as ordered Weber objective functions.

A different approach to handle the nonconvexity of the objective function can be seen in the application of global optimization methods (see, for example Hansen et al., 1995). Krau (1996) and Hansen et al. (2000) generalized the big square small square method, a geometrical branch and bound algorithm suggested by Hansen et al. (1985) and Plastria (1992), to handle polyhedral barrier sets as well as forbidden regions. The big square small square method splits the feasible region into squares, and either rejects a square or further subdivides it based on the evaluation of lower bounds. On the other hand, Fliege (1997) suggested to model the physical barriers by suitable barrier functions (in the sense of nonlinear optimization), an approach that may yield good approximations for a broad class of location problems with barriers, including multifacility problems.

Related results and a short summary on the subject can also be found in the books of Drezner (1995) and Drezner and Hamacher (2002).

In the following chapters we will discuss theoretical results as well as algorithmic approaches for a broad class of location problems with barriers, including, for example, problems with objective functions other than the Weber objective (Chapters 5 and 9), distance functions suitable to approximate road distances (Chapters 8 and 9), problems with multiple objectives

(Chapter 10), and exploiting the special structure of problems with "simple" barriers (Chapters 6 and 7). Parts of these result were also published in Dearing et al. (2002), Hamacher and Klamroth (2000), Klamroth (2001a), Klamroth (2001b), and Klamroth and Wiecek (2002).

3.3   The Visibility Graph

Suppose that the barrier sets in $\mathcal{B}$ are pairwise disjoint, closed, polyhedral sets in the plane $\mathbb{R}^2$. In analogy to Section 2.4 we refer to the sets of extreme points and facets of $\mathcal{B}$ as $\mathcal{P}(\mathcal{B})$ and $\mathcal{F}(\mathcal{B})$, respectively.

In this planar environment, the concept of visibility (see Definition 2.3 on page 18) implies a network structure interrelating the existing facilities and the extreme points of the barriers, which is extremely useful for an efficient evaluation of objective function values of barrier problems.

Definition 3.1. (Aneja and Parlar, 1994; Butt and Cavalier, 1996)
Let d be a metric induced by a norm. A visibility graph of $\mathcal{E}x \cup \mathcal{P}(\mathcal{B})$ is a graph $G_d = (V(G_d), E(G_d))$ with node set $V(G_d) := \mathcal{E}x \cup \mathcal{P}(\mathcal{B})$. Two nodes $v_i, v_j \in V(G_d)$ are connected by an edge $\{v_i, v_j\} \in E(G_d)$ of length $d(v_i, v_j)$ if the corresponding points in the plane are d-visible and have distance $d(v_i, v_j)$.

In Figure 3.1 an example is given for the case that distances are measured by the Manhattan metric l_1 and a single triangular barrier restricts traveling.

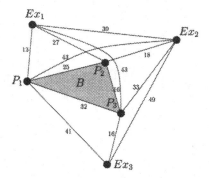

Figure 3.1. The visibility graph for an example problem where distances are measured with respect to the l_1 metric.

Observe that the visibility graph G_d facilitates the determination of the barrier distance between existing facilities and all those points that are candidates for intermediate points on d-shortest permitted paths between the existing facilities and a feasible point $X \in \mathcal{F}$:

Following Corollary 2.3 on page 34, the barrier distance $d_{\mathcal{B}}(Ex_m, X)$ between an existing facility $Ex_m \in \mathcal{E}x$ and a point $X \in \mathcal{F}$ can be calculated as

$$d_B\left(Ex_m, X\right) = d_G\left(Ex_m, I_{Ex_m,X}\right) + d\left(I_{Ex_m,X}, X\right), \qquad (3.8)$$

where $d_G\left(Ex_m, I_{Ex_m,X}\right)$ denotes the length of a shortest path between Ex_m and the intermediate point $I_{Ex_m,X}$ in the visibility graph G_d.

In the case of Euclidean distances $d = l_2$, the visibility graph of N disjoint (not necessarily convex) barriers having a constant number of extreme points (i.e., $|\mathcal{P}(\mathcal{B})| = O(N)$) can be computed in $O\left(N^2\right)$ time (see Alt and Welzl, 1988; Welzl, 1985). An algorithm with an output-sensitive running time of $O(N \log N + |E(G_{l_2})|)$ has been given in Ghosh and Mount (1991), where $|E(G_{l_2})|$ denotes the number of edges in G_{l_2}. For an alternative algorithm with a running time of $O(|E(G_{l_2})| \log N)$ see Overmars and Welzl (1988). An overview of recent results on visibility graphs can be found, for example, in O'Rourke (1993) and Shermer (1992).

4

Bounds for Location Problems with Barriers

Since location problems with barriers are generally difficult global optimization problems, problem relaxations allowing for the determination of good bounds are essential for their solution. Some of the bounds for location problems with barriers developed in this chapter are also discussed in Dearing et al. (2002) and Hamacher and Klamroth (2000).

Throughout this chapter we assume that all barriers in $\mathcal{B} = \bigcup_{i=1}^{N} B_i$ are closed and pairwise disjoint sets in the n-dimensional space $\mathbb{R}^n$. Neither boundedness nor convexity of the barriers will be needed, and whenever the barriers are assumed to be polyhedral sets, this will be indicated. A finite set of existing facilities $\mathcal{E}x = \{Ex_m \in \mathcal{F} : m \in \mathcal{M}\}$, $\mathcal{M} = \{1, \ldots, M\}$ is given in a connected subset of the feasible region $\mathcal{F}$. Furthermore, we assume that a metric d induced by a norm $\| \bullet \|_d$ is given by $d(X, Y) = \|Y - X\|_d$ for $X, Y \in \mathbb{R}^n$.

Using the distance measure $d_{\mathcal{B}}$ as defined in Definition 2.2 on page 17, we consider the following general location problem, classified as $1/\mathbb{R}^n/\mathcal{B}/d_{\mathcal{B}}/f$:

$$\min \quad f_{\mathcal{B}}(X) = f(d_{\mathcal{B}}(X, Ex_1), \ldots, d_{\mathcal{B}}(X, Ex_M))$$

$$\text{s.t.} \quad X \in \mathcal{F}.$$

Here $f : \mathbb{R}^n \to \mathbb{R}$ with $f(X) = f(d(X, Ex_1), \ldots, d(X, Ex_M))$ is the objective function originating from a corresponding unconstrained problem and depending on the distances between the new location X and the existing facilities $Ex_1, \ldots, Ex_M$. Furthermore, we assume that f is a nondecreasing function of the distances $d(X, Ex_m)$, $m \in \mathcal{M}$, such as, for example, the Weber objective function (3.2) or the center objective function (3.3).

Lemma 4.1. *Let f be the objective function of a location problem of type $1/\mathbb{R}^n/\mathcal{B}/d_\mathcal{B}/f$, and suppose that f is a nondecreasing function of the distances $d(X, Ex_m)$, $m \in \mathcal{M}$. Then*

$$f(X) \leq f_\mathcal{B}(X) \qquad \forall X \in \mathcal{F}.$$

Proof. The result follows immediately from Corollary 2.1 on page 18, which implies that $d(X, Ex_m) \leq d_\mathcal{B}(X, Ex_m)$, $m \in \mathcal{M}$, holds for all $X \in \mathcal{F}$. □

4.1   Lower Bounds

Perhaps the easiest way to obtain lower bounds for location problems with barriers is to consider the corresponding unconstrained problems, simply discarding the barrier regions in $\mathcal{B}$. In other words, instead of solving a problem of type $1/\mathbb{R}^n/\mathcal{B}/d_\mathcal{B}/f$ with some objective function $f : \mathbb{R}^n \to \mathbb{R}$ that is a nondecreasing function of the distances $d_\mathcal{B}(X, Ex_m)$, $m \in \mathcal{M}$, we solve the corresponding problem $1/\mathbb{R}^n/ \bullet /d/f$.

Theorem 4.1. *Let $z_\mathcal{B}^*$ be the optimal objective value of a barrier problem of the type $1/\mathbb{R}^n/\mathcal{B}/d_\mathcal{B}/f$ and let X^* be an optimal solution of the corresponding unconstrained problem of type $1/\mathbb{R}^n/ \bullet /d/f$. Then*

$$f(X^*) \leq z_\mathcal{B}^*.$$

Proof. Let $X_\mathcal{B}^*$ be an optimal solution of $1/\mathbb{R}^n/\mathcal{B}/d_\mathcal{B}/f$. Since X^* minimizes f in $\mathbb{R}^n$ and since $f(X) \leq f_\mathcal{B}(X)$ for all $X \in \mathcal{F}$, we obtain

$$f(X^*) \leq f(X_\mathcal{B}^*) \leq f_\mathcal{B}(X_\mathcal{B}^*) = z_\mathcal{B}^*.$$ □

Figure 4.1 gives an example where the bound given in Theorem 4.1 is not sharp.

A drawback for the applicability of Theorem 4.1 is that the optimal solution X^* of the unconstrained problem $1/\mathbb{R}^n/ \bullet /d/f$ may not be feasible for the constrained problem $1/\mathbb{R}^n/\mathcal{B}/d_\mathcal{B}/f$. This can be avoided by considering a different relaxation of $1/\mathbb{R}^n/\mathcal{B}/d_\mathcal{B}/f$ to a restricted location problem involving forbidden regions: While it is still forbidden to place a new facility in the interior of the forbidden region $\mathcal{R} := \mathcal{B}$, trespassing through the interior of $\mathcal{R}$ is allowed. For this problem, classified as $1/\mathbb{R}^n/(\mathcal{R} = \mathcal{B})/d/f$, different solution concepts are available, depending on the dimension n, the objective function f, and the prescribed metric d.

Theorem 4.2. *Let $z_\mathcal{B}^*$ be the optimal objective value of a barrier problem $1/\mathbb{R}^n/ \mathcal{B}/d_\mathcal{B}/f$ and let $X_\mathcal{R}^*$ be an optimal solution of the restricted problem $1/\mathbb{R}^n/(\mathcal{R} = \mathcal{B})/d/f$. Then*

$$f(X_\mathcal{R}^*) \leq z_\mathcal{B}^*.$$

Proof. Let $X_\mathcal{B}^*$ be an optimal solution of the barrier problem. Since $X_\mathcal{R}^*$ is an optimal solution of the restricted problem and since $f(X) \leq f_\mathcal{B}(X)$ holds for all $X \in \mathcal{F}$, we have

$$f(X_\mathcal{R}^*) \leq f(X_\mathcal{B}^*) \leq f_\mathcal{B}(X_\mathcal{B}^*) = z_\mathcal{B}^*. \qquad \square$$

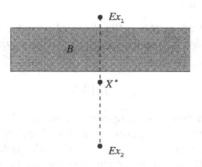

Figure 4.1. A center problem $1/P/\mathcal{B}/l_{2,\mathcal{B}}/$ max with two existing facilities having equal weights. The optimal solution X^* of the corresponding unconstrained problem $1/P/ \bullet /l_2/$ max lies in $\mathcal{F}$ and satisfies $f(X^*) < z_\mathcal{B}^*$. However, X^* is not optimal for $1/P/\mathcal{B}/l_{2,\mathcal{B}}/$ max; i.e., $f_\mathcal{B}(X^*) > z_\mathcal{B}^*$.

An immediate consequence of Theorem 4.2 is the next result.

Corollary 4.1. *Let* $X_\mathcal{R}^*$ *be an optimal solution of the restricted problem* $1/\mathbb{R}^n/\ (\mathcal{R} = \mathcal{B})/d/f$. *If* $d(Ex_i, X_\mathcal{R}^*) = d_\mathcal{B}(Ex_i, X_\mathcal{R}^*)$ *for all* $i = 1, \dots, M$, *then* $X_\mathcal{R}^* = X_\mathcal{B}^*$ *is an optimal solution of* $1/\mathbb{R}^n/\mathcal{B}/d_\mathcal{B}/f$.

Particularly for planar location problems with convex forbidden regions many solution strategies are available that allow the efficient determination of bounds for the corresponding location problems with barriers according to Theorem 4.2.

For the case that distances are measured by a metric induced by a norm, $1/P/\mathcal{R}/d/\sum$, Hamacher (1995) showed that $X_\mathcal{R}^*$ is located on the boundary $\partial\mathcal{R}$ of the forbidden region $\mathcal{R}$ if $X^* \in \text{int}(\mathcal{R})$ holds for the corresponding unrestricted problem.

Theoretical results as well as solution strategies for location problems with forbidden regions can be found in Aneja and Parlar (1994), Batta et al. (1989), Brady and Rosenthal (1980, 1983), Drezner (1983), Eckhardt (1975), Fliege and Nickel (1997), Foulds and Hamacher (1993), Hansen et al. (1982), Hansen et al. (1985), Hamacher (1995), Hamacher and Nickel (1994, 1995, 2001), Hamacher and Schöbel (1997), Karkazis (1988), Nickel (1991, 1994, 1995, 1997, 1998), Nickel and Hamacher (1992), and Plastria (1992), among others.

4.2 Upper Bounds

Since the restricted location problem $1/\mathbb{R}^n/(\mathcal{R} = \mathcal{B})/d/f$ with the forbidden region $\mathcal{R} := \mathcal{B}$ always yields a feasible solution $X_{\mathcal{R}}^*$ of the barrier problem $1/\mathbb{R}^n/\mathcal{B}/d_{\mathcal{B}}/f$, an upper bound for location problems with barriers can be obtained by evaluating the objective value of $X_{\mathcal{R}}^*$ with respect to the barrier objective function $f_{\mathcal{B}}$. Thus the following result, similar to Theorem 4.2, is obtained:

Corollary 4.2. *Let $z_{\mathcal{B}}^*$ be the optimal objective value of a barrier problem $1/\mathbb{R}^n/\,\mathcal{B}/d_{\mathcal{B}}/f$ and let $X_{\mathcal{R}}^*$ be an optimal solution of the restricted problem $1/\mathbb{R}^n/(\mathcal{R} = \mathcal{B})/d/f$. Then*

$$z_{\mathcal{B}}^* \leq f_{\mathcal{B}}(X_{\mathcal{R}}^*).$$

A different approach to deriving upper bounds for location problems with barriers can be applied in the planar case if all barriers are polyhedral sets, i.e., to problems of type $1/P/(\mathcal{B} = N\,polyhedra)/d_{\mathcal{B}}/f$. This approach makes use of the visibility graph of the problem to interrelate barrier problems with network location problems.

Let $\mathcal{B}$ be the union of a finite set of closed polyhedra in the plane $\mathbb{R}^2$. As usual, we denote the set of extreme points of $\mathcal{B}$ by $\mathcal{P}(\mathcal{B})$. Then a node network location problem can be defined on the visibility graph G_d of $\mathcal{E}x \cup \mathcal{P}(\mathcal{B})$ (cf. Definition 3.1 on page 47) in the following way: Each node $v \in V(G_d) = \mathcal{E}x \cup \mathcal{P}(\mathcal{B})$ is interpreted as an existing facility with weight $w(v) = 0$ if $v \in \mathcal{P}(\mathcal{B})$, and as an existing facility with weight $w(v) = w(Ex_m)$ if $v = Ex_m \in \mathcal{E}x$. According to Definition 3.1, every pair of nodes $v_i, v_j \in V(G_d)$ that are d-visible in the embedding of G_d in $\mathcal{F}$ are connected by an edge of length $d(v_i, v_j)$.

The node network location problem $1/G_d/\bullet/d_G(V, V)/f$ on G_d is then given by

$$\min \quad f_G(v)$$

$$\text{s.t.} \quad v \in V(G_d),$$

where $f_G(v) = f(d_G(v, Ex_1), \ldots, d_G(v, Ex_M))$. Note that since the feasible region $\mathcal{F}$ is assumed to be a connected set, Lemma 2.3 on page 31 implies that the problem $1/G_d/\bullet/d_G(V, V)/f$ has at least one feasible solution.

Theorem 4.3. *Let $\mathcal{B}$ be the union of a finite set of closed, polyhedral barriers in $\mathbb{R}^2$ with a finite set of extreme points $\mathcal{P}(\mathcal{B})$. Let $\mathcal{E}x$ be a finite set of existing facilities in $\mathcal{F}$, and let $z_{\mathcal{B}}^*$ be the optimal objective value of a barrier problem $1/P/(\mathcal{B} = N\,polyhedra)/d_{\mathcal{B}}/f$. Furthermore, let G_d be the visibility graph of the existing facilities and the extreme points of the barriers according to Definition 3.1. If X_G^* is an optimal solution of the node network location problem $1/G_d/\bullet/d_G(V, V)/f$ on G_d, then the corresponding point X_G^* of the embedding of G_d in the plane is feasible for*

$1/P/(\mathcal{B} = N\,polyhedra)/d_{\mathcal{B}}/f$ and

$$z_{\mathcal{B}}^* \leq f_G(X_G^*).$$

Proof. The feasibility of X_G^* is trivial because $X_G^* \in V(G_d) = \mathcal{E}x \cup \mathcal{P}(\mathcal{B})$. Using (3.8) and the fact that $d_{\mathcal{B}}(X, Ex_m) = d_G(X, Ex_m)$ for all $X \in \mathcal{E}x \cup \mathcal{P}(\mathcal{B})$ and all $m \in \mathcal{M}$, the upper bound on the optimal objective value of the barrier problem follows from

$$
\begin{aligned}
z_{\mathcal{B}}^* &= \min_{X \in \mathcal{F}} f(d_{\mathcal{B}}(X, Ex_1), \ldots, d_{\mathcal{B}}(X, Ex_M)) \\
&\leq \min_{X \in \mathcal{E}x \cup \mathcal{P}(\mathcal{B})} f(d_{\mathcal{B}}(X, Ex_1), \ldots, d_{\mathcal{B}}(X, Ex_M)) \\
&= \min_{X \in V(G_d)} f(d_G(X, Ex_1), \ldots, d_G(X, Ex_M)) \\
&= f_G(X_G^*). \qquad \square
\end{aligned}
$$

An example for the application of Theorem 4.3 is given in Figure 4.2.

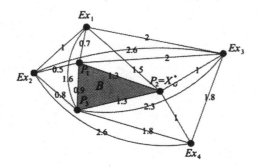

Figure 4.2. The visibility graph G_d for a Weber problem with a block norm. If the weights of all existing facilities are equal to one, the optimal solution of the corresponding node network location problem on G_d is $X_G^* = P_2$ with objective value $f_G(X_G^*) = 5.3$.

Even though the bounds derived above may lead to good approximations or even to optimal solutions in many applications, bounds with a proven quality can in general not be expected. In fact, examples can be constructed for which these bounds become arbitrarily bad with respect to the optimal solution of the problem:

For the upper bound based on the relaxation to restricted location problems $\_/\mathbb{R}^n/(\mathcal{R} = \mathcal{B})/d/\sum$, consider an arbitrarily long but narrow barrier region in the plane $\mathbb{R}^2$ and three existing facilities with equal weights, two of which are located on the same side of the barrier. Obviously, an optimal solution of $1/P/\mathcal{B}/d_{\mathcal{B}}/\sum$ would then be located on the same side of the barrier as the two existing facilities, whereas the barrier may be placed so that the solution of $1/P/(\mathcal{R} = \mathcal{B})/d/\sum$ is located on the opposite side of the barrier. Since increasing the length of the barrier increases the error of the approximation, the corresponding bound may become arbitrarily weak.

For the upper bound based on the visibility graph of the problem, an example with similar properties can be constructed even without introducing a barrier region. Let three existing facilities with equal weights be located on the corner points of an equilateral triangle. Then the optimal solution of $1/P/\bullet/l_2/\max$ is located at the center of the triangle, whereas any solution of $1/G_d/\bullet/d_G(V,V)/\max$ is located in a corner point of the triangle. Increasing the size of the triangle again leads to an arbitrarily bad bound.

However, location problems with barriers are generally hard to solve non-convex optimization problems, and bounds are crucial for the development of efficient solution procedures. Since the bounds given in this section can be easily calculated, their application can significantly facilitate the solution of problems of the type $1/\mathbb{R}^n/\mathcal{B}/d_\mathcal{B}/f$.

Part II

Solution Methods for Specially Shaped Barriers

5

Planar Location Problems with Polyhedral Barriers

In this chapter a strong relationship between location problems with polyhedral barriers and the corresponding unconstrained location problems is developed. See also Klamroth (2001b) for related results.

A physical barrier encountered in real-life problems very often has edges or somewhat regular boundaries allowing a polyhedron to be a fairly good approximation of its shape. Moreover, any given convex set can be approximated arbitrarily well by a convex polyhedron (see Burkard et al., 1991; Valentine, 1964). Polyhedral barriers, on the other hand, have some nice mathematical properties that can be exploited in the context of location theory.

Objective functions will be considered that are convex functions of distances between a set of existing facilities and one new facility. This definition of the objective function includes the Weber objective function and the center objective function as well as ordered Weber objective functions (cf. Section 3.1) as special cases.

Even though parts of the concepts developed in this and in the following chapters can be generalized to higher-dimensional problems, the focus will be on the planar case, since this case seems to be the most relevant in practice. Moreover, some useful properties of shortest paths can be proven only in the 2-dimensional case (cf. Sections 2.3 and 2.4), implying additional features of location problems in the plane $\mathbb{R}^2$ that could not be exploited in the case of higher-dimensional problems.

Throughout this chapter we assume that a finite set of convex, closed, polyhedral and pairwise disjoint barriers $\{B_1, \ldots, B_N\}$ is given in the plane

$\mathbb{R}^2$. As usual, we denote the union of the barrier regions by $\mathcal{B} = \bigcup_{i=1}^{N} B_i$ and the finite sets of extreme points and facets of $\mathcal{B}$ by $\mathcal{P}(\mathcal{B})$ and $\mathcal{F}(\mathcal{B})$, respectively. The connected feasible region $\mathcal{F}$ for new locations is given by $\mathcal{F} = \mathbb{R}^2 \setminus \text{int}(\mathcal{B})$. A finite set of existing facilities $\mathcal{E}x = \{Ex_m \in \mathcal{F} : m \in \mathcal{M}\}$, $\mathcal{M} = \{1, \ldots, M\}$ is given in $\mathcal{F}$. Moreover, a metric d induced by a norm $\| \bullet \|_d$ is given by $d(X, Y) = \|Y - X\|_d$ for all $X, Y \in \mathbb{R}^2$, and the corresponding barrier distance $d_\mathcal{B}$ is defined according to Definition 2.2.

We consider the following general barrier location problem, classified as $1/P/(\mathcal{B} = N \; convex \; polyhedra)/d_\mathcal{B}/f \; convex$:

$$
\begin{aligned}
\min \quad & f_\mathcal{B}(X) = f(d_\mathcal{B}(X, Ex_1), \ldots, d_\mathcal{B}(X, Ex_M)) \\
\text{s.t.} \quad & X \in \mathcal{F},
\end{aligned}
\tag{5.1}
$$

where the objective function $f : \mathbb{R}^n \to \mathbb{R}$ with $f(X) = f(d(X, Ex_1), \ldots, d(X, Ex_M))$ of the corresponding unconstrained problem is assumed to be a convex and nondecreasing function of the distances $d(X, Ex_1), \ldots, d(X, Ex_M)$.

In the following section a reduction result is developed that interrelates location problems with polyhedral barriers with a finite set of unconstrained location problems. In Section 5.2 some theoretical implications of this result are discussed. The algorithmic consequences of the reduction result are derived in Section 5.3, and a mixed-integer programming formulation is suggested in Section 5.4. The case of the Weber objective function is given special attention in Section 5.5, and an extension of the results to multifacility Weber problems is developed in Section 5.6.

5.1   Interrelations Between Barrier Problems and Unconstrained Location Problems

For the special case of the Weber problem with convex polyhedral barriers and the Euclidean distance function, Butt and Cavalier (1996) proposed a decomposition of the feasible region into a finite number of subregions $R_k \subseteq \mathcal{F}$ such that the intermediate points $I_{Ex_m, X}$, $m \in \mathcal{M}$ (cf. Definition 2.5 on page 33), that can be used to evaluate the shortest barrier distance between a point $X \in R_k$ and the existing facilities $Ex_m \in \mathcal{E}x$ remain constant throughout the subregion R_k:

Definition 5.1. (Butt and Cavalier, 1996) *For a finite set of (not necessarily convex) polyhedral barriers in $\mathbb{R}^2$ with extreme points $\mathcal{P}(\mathcal{B})$ and a finite set of existing facilities $\mathcal{E}x$, a region of constant intermediate points $R_k \subseteq \mathcal{F}$ is a nonempty subset of $\mathcal{F}$ such that there exists an assignment of intermediate points $I_m^k \in \mathcal{E}x \cup \mathcal{P}(\mathcal{B})$, $m \in \mathcal{M}$, to the existing facilities in*

$\mathcal{E}x$ with

$$R_k = \{X \in \mathcal{F} : l_{2,\mathcal{B}}(X, Ex_m) = l_2\left(X, I_m^k\right) + l_{2,\mathcal{B}}\left(I_m^k, Ex_m\right) \ \forall m \in \mathcal{M}\}$$
$$\cap \{X \in \mathcal{F} : X \in \text{visible}_{l_2}\left(I_m^k\right) \ \forall m \in \mathcal{M}\}.$$

Definition 5.1 defines a decomposition of the feasible region $\mathcal{F}$ into a finite number of regions of constant intermediate points, since there exist only finitely many different assignments of intermediate points in $\mathcal{E}x \cup \mathcal{P}(\mathcal{B})$ to the existing facilities in $\mathcal{E}x$. In other words, Definition 5.1 implies a subdivision of $\mathcal{F}$ into a finite number of subsets $\{R_1, \dots, R_K\}$ satisfying $\bigcup_{k=1}^{K} R_k = \mathcal{F}$ and $\text{int}(R_i) \cap \text{int}(R_j) = \emptyset$ for all $i, j \in \{1, \dots, K\}$, $i \neq j$. The subsets $R_1, \dots, R_K$ can be obtained by intersecting the Euclidean shortest path maps (SPMs) of all existing facilities (see, for example, Mitchell, 2000, for the construction of shortest path maps).

Using this decomposition of $\mathcal{F}$, an optimal solution of a problem $1/P/$ $(\mathcal{B} = N \, convex \, polyhedra)/l_{2,\mathcal{B}}/\sum$ in the subregion R_k can be found as the optimal solution of an unconstrained problem $1/P/(\mathcal{F} = R_k)/l_2/\sum$, restricted to the region of constant intermediate points R_k, and with existing facilities at the points $I_m^k = I_{Ex_m,X}$, $m \in \mathcal{M}$. This problem is a convex problem on every convex subset of the possibly nonconvex feasible set R_k. Consequently, the original problem $1/P/(\mathcal{B} = N \, convex \, polyhedra)/l_{2,\mathcal{B}}/$ $\sum$ has a convex objective function on every convex subset of a region of constant intermediate points R_k and can be solved by solving the corresponding convex subproblems, at least one subproblem on each of the nonempty subregions R_k.

A drawback of this approach is that the boundaries of the subregions R_k are in general nonlinear, as can be seen in Figure 5.1. In the example of Figure 5.1, the boundary between the two regions of constant intermedi-

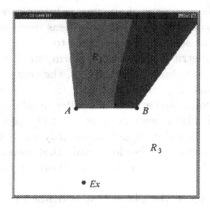

Figure 5.1. If only one existing facility Ex is given and if the barrier is represented by a line segment, we obtain three subregions in which the intermediate points on l_2-shortest paths to Ex remain constant.

ate points R_1 and R_2 in the l_2-shadow of the existing facility Ex can be represented as follows:

Let $A = (a_1, a_2)^T$ and $B = (b_1, b_2)^T$ be the left and right endpoints of the barrier segment, respectively, and let $k_A := l_2(Ex, A)$ and $k_B := l_2(Ex, B)$ denote the distance from the given existing facility to A and B, respectively. Then all points $X = (x_1, x_2)^T$ on the curve separating R_1 and R_2 satisfy

$$k_A + l_2(A, X) = k_B + l_2(B, X). \tag{5.2}$$

Obviously, equation (5.2) defines a line in $\mathbb{R}^2$ if $k_A = k_B$. Otherwise, (5.2) is equivalent to

$$k_A + \sqrt{(a_1 - x_1)^2 + (a_2 - x_2)^2} = k_B + \sqrt{(b_1 - x_1)^2 + (b_2 - x_2)^2}$$

and implicitly defines a smooth nonlinear curve as depicted in Figure 5.1. Note that R_1 is in this case nonconvex.

Especially when the number of barrier regions and existing facilities increases, the determination and representation of the subregions R_k is impractical. To overcome this difficulty, a heuristic method is suggested in Butt and Cavalier (1996) that avoids the explicit calculation of the subregions R_k. Starting with some initial solution $X \in \mathcal{F}$, the procedure iteratively solves unconstrained Weber problems with respect to the intermediate points corresponding to the current solution, subject to some feasibility constraints. This approach is computationally very efficient, and an optimal solution of the problem is found in the majority of cases. However, an optimal solution cannot be guaranteed by this procedure.

A different decomposition of the feasible region is suggested in this section. It avoids the determination of subregions with nonlinear boundaries but still allows the development of an exact solution procedure to solve the nonconvex barrier problem. Moreover, the ideas developed in this chapter are more general in the sense that they apply to all location problems (5.1), with any prescribed metric d induced by a norm, and for a large class of objective functions. This includes, for example, the center objective function and the ordered Weber objective function.

The suggested decomposition uses a smaller number of subregions, and the boundaries of all subregions can be shown to be piecewise linear. Even though the objective function is not necessarily convex on each of these subregions, we will prove a reduction result that nevertheless implies an exact algorithm for the solution of (5.1) based on a grid tessellation of the feasible region.

Consider the grid $\mathcal{G}_d$ in the plane that is defined by the boundaries of the shadows of all existing facilities and of all extreme points of the barrier regions, plus all facets of the barrier regions:

Definition 5.2. *The grid*

$$\mathcal{G}_d := \left(\bigcup_{X \in \mathcal{E}x \cup \mathcal{P}(\mathcal{B})} \partial(\mathrm{shadow}_d(X)) \right) \cup \mathcal{F}(\mathcal{B})$$

is called the visibility grid *with respect to $\mathcal{E}x$ and $\mathcal{B}$.*

The set of cells *of $\mathcal{G}_d$, i.e., the set of all polyhedra in $\mathcal{F}$ induced by $\mathcal{G}_d$ the nonempty interior of which is not intersected by a line segment in $\mathcal{G}_d$ is denoted by $\mathcal{C}(\mathcal{G}_d)$.*

Since the barriers are convex polyhedra and since the boundary of $\mathrm{shadow}_d(X)$ is piecewise linear for all $X \in \mathcal{F}$ (cf. Lemma 2.6 on page 35), the grid $\mathcal{G}_d$ always consists of a finite set of line segments in $\mathcal{F}$, as can be seen in Figure 5.2.

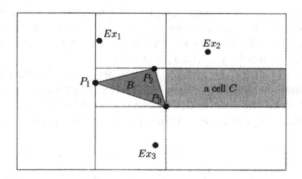

Figure 5.2. The grid $\mathcal{G}_{l_1}$ for an example problem with one triangular barrier and three existing facilities.

Consequently, the visibility grid $\mathcal{G}_{l_2}$ for the example given in Figure 5.1 does not contain the nonlinear curve separating R_1 and R_2 and implies a tessellation of the feasible region into only the two cells $C_1 = R_1 \cup R_2$ and $C_2 = R_3$.

Even though all cells in $\mathcal{C}(\mathcal{G}_d)$ are polyhedral sets (cf. Lemma 2.6), they are not necessarily convex, as can be seen in Figure 5.3.

The size of the visibility grid $\mathcal{G}_d$ depends on d, on the number of existing facilities $|\mathcal{E}x|$, and on the number of extreme points of the barrier regions $|\mathcal{P}(\mathcal{B})|$. In the case that $d = l_2$, i.e., in the case of Euclidean distances, the number of half-lines defining $\mathcal{G}_{l_2}$ is bounded by $O((|\mathcal{E}x| + |\mathcal{P}(\mathcal{B})|) \cdot |\mathcal{P}(\mathcal{B})|)$. This implies at most $O\left((|\mathcal{E}x| + |\mathcal{P}(\mathcal{B})|)^2 \cdot |\mathcal{P}(\mathcal{B})|^2\right)$ intersection points. If we interpret $\mathcal{G}_{l_2}$ as a planar graph with a vertex at each intersection point, this graph has at most $O\left((|\mathcal{E}x| + |\mathcal{P}(\mathcal{B})|)^2 \cdot |\mathcal{P}(\mathcal{B})|^2\right)$ edges, since every half-line in $\mathcal{G}_{l_2}$ can be intersected by at most $O((|\mathcal{E}x| + |\mathcal{P}(\mathcal{B})|) \cdot |\mathcal{P}(\mathcal{B})|)$ other half-lines, and thus these contribute at most $O((|\mathcal{E}x| + |\mathcal{P}(\mathcal{B})|) \cdot |\mathcal{P}(\mathcal{B})|)$ edges to the graph. The number of cells in $\mathcal{C}(\mathcal{G}_{l_2})$ can now be estimated

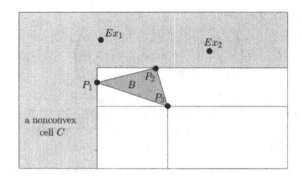

Figure 5.3. The grid $\mathcal{G}_{l_1}$ for the example problem of Figure 5.2, but without the existing facility Ex_3.

using Euler's formula for planar graphs: The number of vertices minus the number of edges plus the number of cells of every planar graph equals 2 (see, for example, Harary, 1969). This implies that the number of cells in $\mathcal{C}(\mathcal{G}_{l_2})$ is also bounded by $O\left((|\mathcal{E}x| + |\mathcal{P}(\mathcal{B})|)^2 \cdot |\mathcal{P}(\mathcal{B})|^2\right)$.

Moreover, all cells of the grid $\mathcal{G}_{l_2}$ are convex polyhedra (see Figure 5.4), since we assumed that the barriers are convex polyhedra that are pairwise disjoint and since the cells in $\mathcal{C}(\mathcal{G}_{l_2})$ are bounded by facets of the barriers or by half-lines originating in extreme points of the barrier polyhedra.

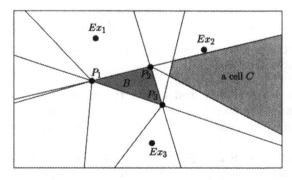

Figure 5.4. The grid $\mathcal{G}_{l_2}$ for the example problem introduced in Figure 5.2.

In contrast to the definition of the visibility graph in Section 3.3, the visibility grid does not directly supply information about the shortest barrier distance between particular points in the plane. It decomposes the feasible region $\mathcal{F}$ into a finite number of subregions (cells) so that candidates for intermediate points on d-shortest permitted paths to the existing facilities are d-visible from a complete cell C, or they may not be d-visible from the cell C at all.

Lemma 5.1. *Let $C \in \mathcal{C}(\mathcal{G}_d)$ be a cell, let $X \in \text{int}(C)$ be a feasible solution of $1/P/(\mathcal{B} = N$ convex polyhedra$)/d_{\mathcal{B}}/f$ convex, and let $I \in \mathcal{E}x \cup \mathcal{P}(\mathcal{B})$. If I is d-visible from X, then I is d-visible from all points in the cell C.*

Proof. Suppose that $I \in \mathcal{E}x \cup \mathcal{P}(\mathcal{B})$ is d-visible from $X \in \text{int}(C)$. According to Definition 5.2, the interior of C is not intersected by $\partial(\text{shadow}_d(I))$, and therefore either $\text{int}(C) \subseteq \text{shadow}_d(I)$ or $C \subseteq \text{visible}_d(I)$. Since I is d-visible from X we can conclude that $C \subseteq \text{visible}_d(I)$. $\square$

Lemma 5.2. *Let $C \in \mathcal{C}(\mathcal{G}_d)$ be a cell, let $X \in C$ be a feasible solution of $1/P/(\mathcal{B} = N$ convex polyhedra$)/d_{\mathcal{B}}/f$ convex, and let $Ex \in \mathcal{E}x$. Then there exists an intermediate point $I = I_{Ex,X}$ on a d-shortest permitted Ex-X path that is d-visible from all points in C.*

Proof. If $X \in \text{int}(C)$, the result follows immediately from Lemma 5.1. Therefore, let $X \in \partial(C)$ and let $\bar{I} \in \mathcal{E}x \cup \mathcal{P}(\mathcal{B})$ be an intermediate point on a d-shortest permitted Ex-X path SP, satisfying the barrier touching property, that is not only d-visible from X, but that is also l_2-visible from X (cf. Lemma 2.3 on page 31). If $\bar{I}$ is d-visible from some point in $\text{int}(C)$, Lemma 5.1 implies that $I := \bar{I}$ has the desired property.

Otherwise, $\bar{I}$ is neither d-visible nor l_2-visible from any point in $\text{int}(C)$, i.e.,

$$\text{int}(C) \subseteq \text{shadow}_d\left(\bar{I}\right) \subseteq \text{shadow}_{l_2}\left(\bar{I}\right).$$

Since X is l_2-visible from I, X lies on $\partial\left(\text{shadow}_{l_2}\left(\bar{I}\right)\right)$. Thus X lies on a tangent line from $\bar{I}$ to one of the polyhedral barrier sets $B \in \mathcal{B}$, intersecting with B in an extreme point $I \in \mathcal{P}(B)$. This intersection point can be chosen such that it is l_2-visible in the intersection set of an open neighborhood $N_\varepsilon(X) = \{Z \in \mathbb{R}^2 : l_2(Z, X) < \varepsilon\}$ with the cell C for some $\varepsilon > 0$. Note that the existence of a point I with this property follows from the assumption that the number of extreme points of barriers is finite. We can conclude that I is l_2-visible, and therefore also d-visible, from some points in $\text{int}(C) \cap N_\varepsilon(X)$, and therefore Lemma 5.1 implies that I is d-visible from all points in C. Moreover, I lies on SP, since I lies on the straight line segment connecting X and $\bar{I}$. According to Definition 2.5 we can therefore interpret I as an intermediate point on the path SP. $\square$

As with to the representation of the barrier distance with respect to intermediate points in Corollary 2.3 on page 34, the objective function $f_{\mathcal{B}}(X)$ can be rewritten for every point X in a cell $C \in \mathcal{C}(\mathcal{G}_d)$.

Definition 5.3. *Let $C \in \mathcal{C}(\mathcal{G}_d)$ be a cell and let $X \in C$ be a feasible solution of $1/P/(\mathcal{B} = N$ convex polyhedra$)/d_{\mathcal{B}}/f$ convex. Then a function $F_X : \mathbb{R}^2 \to \mathbb{R}$ is defined as*

$$F_X(Y) := f(d(Y, I_1) + c_1, \ldots, d(Y, I_M) + c_M), \quad Y \in \mathbb{R}^2, \tag{5.3}$$

where

$$c_m := d_{\mathcal{B}}(I_m, Ex_m), \quad m = 1, \ldots, M,$$

and where $I_m := I_{Ex_m, X}$, $m = 1, \ldots, M$, *is an intermediate point on a d-shortest permitted* Ex_m-X *path according to Definition 2.5 that is chosen such that it is d-visible from all points in* C.

Note that $d_{\mathcal{B}}(I_m, Ex_m) = d_G(I_m, Ex_m)$ for all $m \in \mathcal{M}$, where $d_G(I_m, Ex_m)$ denotes the network distance between I_m and Ex_m in the visibility graph of $Ex \cup \mathcal{P}(\mathcal{B})$.

Lemma 5.3. *Let* $C \in \mathcal{C}(\mathcal{G}_d)$ *be a cell and let* $X \in C$ *be a feasible solution of* $1/P/(\mathcal{B} = N$ *convex polyhedra)/$d_{\mathcal{B}}/f$ convex. Then*

$$f_{\mathcal{B}}(X) = F_X(X). \tag{5.4}$$

where $F_X : \mathbb{R}^2 \to \mathbb{R}$ *is defined according to Definition 5.3.*

Proof. Let $C \in \mathcal{C}(\mathcal{G}_d)$ be a cell and let $X \in C$. According to Lemma 2.3, and using Lemmas 5.1 and 5.2. there always exist intermediate points $I_m := I_{Ex_m, X}$, $m = 1, \ldots, M$, on d-shortest permitted Ex_m-X paths, satisfying the barrier touching property, that are d-visible from all points in C. Observe that Definition 2.5 implies that $I_m = I_{Ex_m, X} \neq X$ whenever $Ex_m \neq X$. We can conclude that

$$\begin{aligned}
f_{\mathcal{B}}(X) &= f(d_{\mathcal{B}}(X, Ex_1), \ldots, d_{\mathcal{B}}(X, Ex_M)) \\
&= f(d(X, I_1) + d_{\mathcal{B}}(I_1, Ex_1), \ldots, d(X, I_M) + d_{\mathcal{B}}(I_M, Ex_M)) \\
&= F_X(X). \qquad \square
\end{aligned}$$

Observe that the function F_X does not explicitly depend on the nonconvex distance function $d_{\mathcal{B}}$, since the values of $c_1 = d_{\mathcal{B}}(I_1, Ex_1), \ldots, c_M = d_{\mathcal{B}}(I_M, Ex_M)$ are constants not depending on the choice of the argument of F_X. Moreover, $F_X(Y)$ is a convex function of Y, as the following lemma shows:

Lemma 5.4. *If* $f : \mathbb{R}^2 \to \mathbb{R}$ *with* $f(X) = f(d(X, Ex_1), \ldots, d(X, Ex_M))$ *is a convex, nondecreasing function of the distances* $d(X, Ex_1), \ldots, d(X, Ex_M)$, *then the function* $F_X : \mathbb{R}^2 \to \mathbb{R}$ *as defined in Definition 5.3 is a convex function in* $\mathbb{R}^2$.

Proof. Lemma 5.3 and the definition of F_X in Definition 5.3 imply that

$$F_X(Y) = f(d(Y, I_1) + c_1, \ldots, d(Y, I_M) + c_M) = f(\phi_1, \ldots, \phi_M),$$

where $c_m = d_{\mathcal{B}}(I_m, Ex_m)$ are constants. $m = 1, \ldots, M$, and thus $\phi_m : \mathbb{R}^2 \to \mathbb{R}$ with $\phi_m(Y) = d(Y, I_m) + c_m$ are convex functions for all $m = 1, \ldots, M$.

Hence F_X can be interpreted as the composition of the convex, nondecreasing function f and the convex functions $\phi_1, \ldots, \phi_M$, which implies that F_X is also convex (see, for example, Rockafellar and Wets, 1998). $\square$

For two points X and Y in the same cell $C \in \mathcal{C}(\mathcal{G}_d)$, the functions F_X and F_Y relate in the following way:

Lemma 5.5. *Let $C \in \mathcal{C}(\mathcal{G}_d)$ be a cell and let $X \in C$. Then*

$$F_X(Y) \geq F_Y(Y) \qquad \forall Y \in C, \tag{5.5}$$

where F_X and F_Y are defined according to Definition 5.3.

Proof. Let $F_X(Y) = f(d(Y, I_1) + c_1, \ldots, d(Y, I_M) + c_M)$, where $c_m = d_{\mathcal{B}}(I_m, Ex_m)$ and the intermediate points $I_m = I_{Ex_m, X}$ are chosen such that they are d-visible from all points in C, $m = 1, \ldots, M$. Due to the triangle inequality, $d(Y, I_m) + c_m = d_{\mathcal{B}}(Y, I_m) + d_{\mathcal{B}}(I_m, Ex_M) \geq d_{\mathcal{B}}(Y, Ex_m)$ holds for all $m \in \mathcal{M}$ and $Y \in C$. Then the monotonicity of f implies that

$$
\begin{aligned}
F_X(Y) &= f(d(Y, I_1) + c_1, \ldots, d(Y, I_M) + c_M) \\
&\geq f(d_{\mathcal{B}}(Y, Ex_1), \ldots, d_{\mathcal{B}}(Y, Ex_M)) \\
&= F_Y(Y). \qquad \qquad \square
\end{aligned}
$$

Note that Lemma 5.5 is generally not true for points in different cells $C^1, C^2 \in \mathcal{C}(\mathcal{G}_d)$, since in this case the visibility of the intermediate points selected for a point $X \in C^1$ cannot be guaranteed for a point $Y \in C^2$.

The reformulation of the objective function $f_{\mathcal{B}}$ given in Lemma 5.3 will be used in the following to interrelate the nonconvex problem $1/P/(\mathcal{B} = N\ convex\ polyhedra)/d_{\mathcal{B}}/f\ convex$ to a finite set of restricted location problems of type $1/P/(\mathcal{R} = C)/d/F_X$ in each cell $C \in \mathcal{C}(\mathcal{G}_d)$.

Theorem 5.1. *Let $C \in \mathcal{C}(\mathcal{G}_d)$ be a cell and let $X_{\mathcal{B}}^* \in C$ be an optimal solution of $1/P/(\mathcal{B} = N\ convex\ polyhedra)/d_{\mathcal{B}}/f\ convex$. Then $X_{\mathcal{B}}^*$ is an optimal solution of the corresponding restricted location problem*

$$
\begin{aligned}
\min \quad & F_{X_{\mathcal{B}}^*}(Y) \\
\text{s.t.} \quad & Y \in C,
\end{aligned}
\tag{5.6}
$$

where $F_{X_{\mathcal{B}}^*}(Y)$ is defined according to Definition 5.3.

Proof. Let $X_{\mathcal{B}}^* \in C$, let $F_{X_{\mathcal{B}}^*}$ be defined according to Definition 5.3, and let I_m, $m \in \mathcal{M}$ be the corresponding intermediate points on d-shortest permitted Ex_m-$X_{\mathcal{B}}^*$ paths, satisfying the barrier touching property, that are d-visible from all points in C. Lemma 5.5 implies that

$$F_{X_{\mathcal{B}}^*}(Y) \geq F_Y(Y) = f_{\mathcal{B}}(Y) \tag{5.7}$$

holds for all $Y \in C$. Using Lemma 5.3 and the assumption that $X_{\mathcal{B}}^*$ is an optimal solution of $1/P/(\mathcal{B} = N\ convex\ polyhedra)/d_{\mathcal{B}}/f\ convex$, we obtain

$$F_{X_{\mathcal{B}}^*}(Y) \geq f_{\mathcal{B}}(Y) \geq f_{\mathcal{B}}(X_{\mathcal{B}}^*) = F_{X_{\mathcal{B}}^*}(X_{\mathcal{B}}^*) \qquad \forall Y \in C. \qquad \square$$

Theorem 5.1 implies that any problem of type $1/P/(\mathcal{B} = N\ convex$ $polyhedra)/d_\mathcal{B}/f\ convex$ can be reduced to a finite set of convex subproblems within each cell in $\mathcal{C}(\mathcal{G}_d)$ even though the original objective function $f_\mathcal{B}(X)$ is in general nonconvex within the cells. The algorithmic implications of this result will be discussed in Section 5.3.

Note that Theorem 5.1 can be generalized to the case that the given objective function f is nonconvex. Nevertheless, in this case the resulting subproblems are in general also nonconvex, and the difficulty of the problem is not reduced as in the case where f is convex.

If an optimal solution $X_\mathcal{B}^*$ of $1/P/(\mathcal{B} = N\ convex\ polyhedra/d_\mathcal{B}/f\ con$-$vex$ is located in the interior of a cell, the following result proves that this solution can be found by solving a finite set of unconstrained, convex problems of type $1/P/\bullet/d/F_X$ with objective function F_X defined according to Definition 5.3.

Theorem 5.2. *Let $C \in \mathcal{C}(\mathcal{G}_d)$ be a cell and let $X_\mathcal{B}^* \notin \mathcal{G}_d$ be an optimal solution of $1/P/(\mathcal{B} = N\ convex\ polyhedra)/d_\mathcal{B}/f\ convex$. Then $X_\mathcal{B}^*$ is an optimal solution of the corresponding unconstrained, convex problem*

$$\begin{aligned} \min \quad & F_{X_\mathcal{B}^*}(Y) \\ \text{s.t.} \quad & Y \in \mathbb{R}^2. \end{aligned} \tag{5.8}$$

where $F_{X_\mathcal{B}^*}(Y)$ is defined according to Definition 5.3.

Proof. Let $C \in \mathcal{C}(\mathcal{G}_d)$ be a cell such that $X_\mathcal{B}^* \in int(C)$. Since $X_\mathcal{B}^* \in C$, Theorem 5.1 implies that $X_\mathcal{B}^*$ minimizes $F_{X_\mathcal{B}^*}$ in the cell C. Using the fact that $F_{X_\mathcal{B}^*}(Y)$ is a convex function of Y in $\mathbb{R}^2$ and that $X_\mathcal{B}^* \in int(C)$, we can conclude that $X_\mathcal{B}^*$ also minimizes $F_{X_\mathcal{B}^*}(Y)$ in $\mathbb{R}^2$.   □

In some applications it may be beneficial to consider the grid $\mathcal{G}_{l_2}$ instead of the grid $\mathcal{G}_d$ for a given distance function d, especially in the case that the construction of $\mathcal{G}_{l_2}$ is simpler than that of $\mathcal{G}_d$. This is possible for any metric d induced by a norm, since Lemmas 2.2 and 2.3 imply the following reformulation of Theorems 5.1 and 5.2:

Corollary 5.1. *Let d be a metric induced by a norm. Furthermore, let $C \in \mathcal{C}(\mathcal{G}_{l_2})$ be a cell in the grid $\mathcal{G}_{l_2}$ and let $X_\mathcal{B}^* \in C$ be an optimal solution of $1/P/(\mathcal{B} = N\ convex\ polyhedra)/d_\mathcal{B}/f\ convex$. Then $X_\mathcal{B}^*$ is an optimal solution of the corresponding restricted problem*

$$\begin{aligned} \min \quad & F_{X_\mathcal{B}^*}(Y) \\ \text{s.t.} \quad & Y \in C, \end{aligned} \tag{5.9}$$

where $F_{X_\mathcal{B}^*}(Y)$ is defined as

$$F_{X_\mathcal{B}^*}(Y) = f(d(Y, I_1) + c_1, \ldots, d(Y, I_M) + c_M), \quad Y \in \mathbb{R}^2,$$

with $c_m = d_{\mathcal{B}}(I_m, Ex_m)$, $m \in \mathcal{M}$, and where the intermediate points $I_m = I_{Ex_m, X}$, $m \in \mathcal{M}$, on d-shortest permitted $X_{\mathcal{B}}^*$-Ex_m paths with the barrier touching property are chosen such that they are l_2-visible from all points in C (cf. Definition 5.3).

Moreover, if $X_{\mathcal{B}}^* \in \text{int}(C)$, then $X_{\mathcal{B}}^*$ is an optimal solution of the corresponding unconstrained problem

$$\min \quad F_{X_{\mathcal{B}}^*}(Y)$$
$$\text{s.t.} \quad Y \in \mathbb{R}^2. \tag{5.10}$$

Proof. The existence of suitable intermediate points $I_m = I_{Ex_m, X}$ ($m \in \mathcal{M}$) follows from Lemmas 2.2 and 2.3 and from the fact that since the grid $\mathcal{G}_{l_2}$ is used, Lemmas 5.1 and 5.2 can be adopted to l_2-visibility. The rest of the result follows in complete analogy to the proofs of Theorems 5.1 and 5.2. $\qquad\square$

5.2   The Iterative Convex Hull

Using Theorems 5.1 and 5.2, some of the general properties of unconstrained location problems can be transferred to barrier problems $1/P/(\mathcal{B} = N \text{ convex polyhedra})/d_{\mathcal{B}}/f \text{ convex}$.

As an example, in this section location problems are considered for which the set of optimal solutions lies within the convex hull of the existing facilities in the unconstrained case. Defining an *iterative convex hull* $H_{\mathcal{B}}$ of the existing facilities and the barrier regions, an analogous property can be proven for the corresponding barrier problems. However, a nontrivial result can be guaranteed only if we assume the barrier regions to be bounded and thus compact.

Definition 5.4. *Let $\mathcal{B}$ be the union of a finite set of compact, convex, polyhedral and pairwise disjoint barriers in $\mathbb{R}^2$. The* iterative convex hull *$H_{\mathcal{B}}$ of the existing facilities and the barrier regions is defined as the smallest convex set in $\mathbb{R}^2$ such that*

$$\mathcal{E}x \subset H_{\mathcal{B}} \qquad and \qquad \partial H_{\mathcal{B}} \cap \text{int}(\mathcal{B}) = \emptyset.$$

The iterative convex hull $H_{\mathcal{B}}$ of Definition 5.4 can be determined by the following straightforward algorithm:

Algorithm 5.1. (Constructing the Iterative Convex Hull)

Input: *Location problem $1/P/(\mathcal{B} = N \text{ convex polyhedra})/d_{\mathcal{B}}/f \text{ convex}$.*

Step 1:   Set $H := \text{conv}(\mathcal{E}x)$.

Step 2:   While there exists a barrier $B_i \in \mathcal{B}$ such that $\partial H \cap \text{int}(B_i) \neq \emptyset$, set $H := \text{conv}(H, B_i)$.

Output: $H_B := H$.

Figure 5.5 illustrates the application of Algorithm 5.1 to an example with three existing facilities and three barriers.

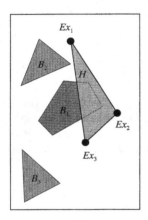 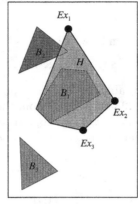 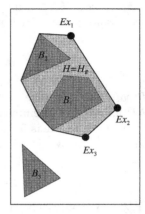

Figure 5.5. Construction of H_B using Algorithm 5.1.

Lemma 5.6. *Algorithm 5.1 determines the iterative convex hull of a finite set of compact, convex, polyhedral and pairwise disjoint barriers in $\mathbb{R}^2$ after a finite number of iterations.*

Proof. Algorithm 5.1 is finite, since we assumed that the number of barrier regions is finite and that all the barrier regions are compact.

To prove the correctness of Algorithm 5.1, let H_i, $i = 1, \ldots, n$, denote the set H after iteration i of the algorithm; i.e., $H_1 := \text{conv}(\mathcal{E}x)$ and H_n is the set constructed at termination of the algorithm. Thus Algorithm 5.1 constructs the iterative convex hull H_B if $H_n = H_B$.

Obviously, $H_i \subseteq H_{i+1}$ holds for all iterations $i \in \{1, \ldots, n-1\}$. Therefore, $\mathcal{E}x \subset H_n$ and, since the boundary of H_n is not intersected by $\text{int}(\mathcal{B})$, it follows that $H_B \subseteq H_n$.

On the other hand, $H_1 \subseteq H_B$ holds after Step 1 of the algorithm. Now assume that $H_n \nsubseteq H_B$. Then there exists an index $i \in \{2, \ldots, n\}$ such that $H_j \subseteq H_B$ for all $j \in \{1, \ldots, i-1\}$, but $H_i \nsubseteq H_B$. Let $B_i \in \mathcal{B}$ be the barrier added in iteration i; i.e., $H_i = \text{conv}(H_{i-1} \cup B_i)$. Consequently, $B_i \nsubseteq H_B$, and since $\partial H_B \cap \text{int}(B_i) = \emptyset$, we have $\text{int}(B_i) \cap H_B = \emptyset$. Using the fact that $H_{i-1} \subseteq H_B$ we can conclude that $\text{int}(B_i) \cap H_{i-1} = \emptyset$, contradicting the construction of the set H_i in iteration i.   □

An algorithm similar to Algorithm 5.1 has been developed in Verbarg (1996) in the context of the determination of Euclidean shortest paths between two points in the presence of barriers. In this algorithm obstacles are iteratively added as soon as their relevance for the determination of a shortest path is detected, a condition that is closely related to the condition that no barrier be intersected by the boundary of $\partial(H_\mathcal{B})$. In Verbarg (1996) an example is given where only one of the N barrier sets is added in each iteration, implying that a total number of $O(N)$ iterations in Algorithm 5.1 may be necessary in the worst case. For an overview of the construction of the convex hull of a set of points or polyhedral sets we refer to Edelsbrunner (1987) and O'Rourke (1994).

Theorem 5.3. *Let $X_\mathcal{B}^* \notin \mathcal{G}_d$ be an optimal solution of $1/P/(\mathcal{B} = N$ convex polyhedra$)/d_\mathcal{B}/f$ convex.*

If for all corresponding unconstrained problems $1/P/\bullet/d/F_X$ with objective function F_X as introduced in Definition 5.3 the set of optimal solutions is contained in the convex hull of the existing facilities, then

$$X_\mathcal{B}^* \in (H_\mathcal{B} \cap \mathcal{F}).$$

Proof. Let $X_\mathcal{B}^*$ be an optimal solution of $1/P/(\mathcal{B} = N$ convex polyhedra$)/d_\mathcal{B}/f$ convex such that $X_\mathcal{B}^* \in \text{int}(C)$ for some cell $C \in \mathcal{C}(\mathcal{G}_d)$.

Suppose that $X_\mathcal{B}^* \notin H_\mathcal{B}$. We can assume that there exists no barrier in $\mathbb{R}^2 \setminus H_\mathcal{B}$, since this assumption does not increase the objective value of $X_\mathcal{B}^*$ and since it does not affect the objective value of any point $X \in (H_\mathcal{B} \cap \mathcal{F})$.

Theorem 5.2 implies that $X_\mathcal{B}^*$ is an optimal solution of problem (5.8) with respect to some intermediate points $I_m \in \mathcal{E}x \cup \mathcal{P}(\mathcal{B})$, $m \in \mathcal{M}$. This problem is an unconstrained location problem of type $1/P/\bullet/d/F_X$ and thus $X_\mathcal{B}^* \in \text{conv}\{I_m : m \in \mathcal{M}\} \cap \mathcal{F}$. Since $H_\mathcal{B}$ is the convex hull of all existing facilities and all barrier sets, we can conclude that

$$\text{conv}\{I_m : m \in \mathcal{M}\} \cap \mathcal{F} \subseteq \text{conv}(\mathcal{E}x \cup \mathcal{P}(\mathcal{B})) \cap \mathcal{F} \subseteq H_\mathcal{B} \cap \mathcal{F}. \qquad \square$$

The assumption of Theorem 5.3 that the set of optimal solutions is contained in the convex hull of the existing facilities for the corresponding unconstrained problem $1/P/\bullet/d/F_X$ is, for example, satisfied for Weber problems, where distances are measured by any l_p metric, $p \in (1, \infty)$; see, for example, Juel and Love (1983). Similar results on the location of the set of optimal solutions of different types of continuous location problems can be found, for example, in Drezner and Goldman (1991), Durier and Michelot (1994), Hansen et al. (1980), Pelegrin et al. (1985), Plastria (1984), Wendell and Hurter (1973), and White (1982).

A similar result can also be proven for location problems for which the set of optimal solutions lies within the rectangular hull of the existing facilities in the unconstrained case, i.e., in the smallest axes-parallel rectangle circumscribing all the existing facilities. This is, for example, satisfied for Weber problems with the l_1 metric $1/P/\bullet/l_1/\sum$; see Love and Morris (1975). Defining the *iterative rectangular hull* of the existing facilities and the barrier regions, analogous to Definition 5.4, as the smallest axes-parallel rectangle $R_\mathcal{B}$ containing all existing facilities and such that $\partial R_\mathcal{B} \cap \text{int}(\mathcal{B}) = \emptyset$, the following result can be proven absolutely analogously to the proof of Theorem 5.3 (see also Theorem 9.1 on page 138):

Corollary 5.2. *Let $X_{\mathcal{B}}^* \notin \mathcal{G}_d$ be an optimal solution of $1/P/(\mathcal{B} = N$ convex polyhedra$)/d_{\mathcal{B}}/f$ convex.*

If for all corresponding unconstrained problems $1/P/\bullet/d/F_X$ with objective function F_X as introduced in Definition 5.3 the set of optimal solutions is contained in the rectangular hull of the existing facilities, then

$$X_{\mathcal{B}}^* \in (R_{\mathcal{B}} \cap \mathcal{F}),$$

where $R_{\mathcal{B}}$ denotes the iterative rectangular hull of $\mathcal{E}x$ and $\mathcal{B}$.

5.3 Algorithmic Consequences

Two different cases may occur when a problem of type $1/P/(\mathcal{B} = N$ convex polyhedra$)/d_{\mathcal{B}}/f$ convex is reduced to a finite set of convex subproblems according to Theorem 5.2 (or Corollary 5.1). An optimal solution $X_{\mathcal{B}}^*$ of a problem of type $1/P/(\mathcal{B} = N$ convex polyhedra$)/d_{\mathcal{B}}/f$ convex may be located either on the grid $\mathcal{G}_d$ or in the interior of a cell $C \in \mathcal{C}(\mathcal{G}_d)$. In the first case $X_{\mathcal{B}}^*$ can easily be found by applying a line search procedure on the line segments of $\mathcal{G}_d$. In the latter case $X_{\mathcal{B}}^*$ is the optimal solution of a corresponding unconstrained problem (5.8).

Hence, a two-step algorithm is suggested to solve problems of the type $1/P/(\mathcal{B} = N$ convex polyhedra$)/d_{\mathcal{B}}/f$ convex. First, a line search procedure is applied on each line segment of the grid $\mathcal{G}_d$. In a second step, a candidate for an optimal solution is sought in the interior of cells in $\mathcal{F} \setminus \mathcal{G}_d$ by solving the convex subproblems (5.8) for all feasible reformulations $f_{\mathcal{B}}(Y) = F_X(Y)$ of the objective function. For each solution Y^* of one of these subproblems feasibility has to be tested; i.e., it has to be verified whether $f_{\mathcal{B}}(Y^*) = F_X(Y^*)$. In the case that the outcome of this feasibility test is negative, Theorem 5.2 implies that the solution Y^* can be discarded from further investigations.

Algorithm 5.2. (Exact Solution of $1/P/(\mathcal{B}=N$ conv. pol.$)/d_{\mathcal{B}}/f$ conv.)

Input: *Location problem $1/P/(\mathcal{B} = N$ convex polyhedra$)/d_{\mathcal{B}}/f$ convex.*

Step 1: *Construct the grid $\mathcal{G}_d$.*

Step 2: *Find the minima of $1/P/(\mathcal{B} = N$ convex polyhedra$)/d_{\mathcal{B}}/ f$ convex on $\mathcal{G}_d$.*

Step 3: *For all feasible reformulations of the objective function; i.e., for all feasible assignments of intermediate points to existing facilities, do:*

(a) *Find an optimal solution Y^* of the corresponding uncon-strained problem* $\min\{F_X(Y) : Y \in \mathbb{R}^2\}$.

(b) *If $f_B(Y^*) = F_X(Y^*)$, the solution Y^* is a candidate for an optimal solution.*

Step 4:   *Determine a global minimum from the candidate set found in Steps 2 and 3.*

Output: *Set of optimal solutions of* $1/P/(B = N$ *conv. pol.)$/d_B/f$ conv.*

Note that the grid $\mathcal{G}_d$ in Algorithm 5.2 can be replaced by the grid $\mathcal{G}_{l_2}$ for any metric d induced by a norm (cf. Corollary 5.1).

The time complexity of Steps 1 and 2 of Algorithm 5.2 depends on the size of the grid $\mathcal{G}_d$ (or $\mathcal{G}_{l_2}$, respectively) and thus on the number of existing facilities, the number of extreme points of the barrier regions, and the choice of the distance function d. In the case that distances are measured by the Euclidean metric l_2, the number of intersection points in $\mathcal{G}_{l_2}$ is bounded by $O\left((|\mathcal{E}x| + |\mathcal{P}(B)|)^2 \cdot |\mathcal{P}(B)|^2\right)$, and the number of cells in $\mathcal{G}_{l_2}$ is bounded by $O\left((|\mathcal{E}x| + |\mathcal{P}(B)|)^2 \cdot |\mathcal{P}(B)|^2\right)$; see also the discussion concerning the grid $\mathcal{G}_{l_2}$ on page 61.

The overall time complexity of Algorithm 5.2 is in general dominated by Step 3. If no additional information is available to reduce the number of assignments of existing facilities to intermediate points, the number of subproblems is exponential in the number of existing facilities and in the number of extreme points of the barrier regions. For specific barrier shapes and distance functions better results are available; see, for example, Chapter 7. However, the strength of Algorithm 5.2 can be seen in its generality and in its applicability to a broad class of different location problems.

A slight modification of this algorithm was implemented based on the Library of Location Algorithms LoLA (see Hamacher et al. (1999a) and Ochs (1998)) for the Euclidean distance function and one circular or poly-hedral barrier. In the case of a circular barrier, tangents to the circle are used to define the boundary of the shadow of the barrier; see Chapter 6 for a more detailed discussion. In this implementation all the convex subprob-lems are solved by adapting the method of Hooke and Jeeves (see Hooke and Jeeves, 1961). Using this implementation it was possible to improve on earlier results obtained by Butt and Cavalier (1996) and Katz and Cooper (1981) for the special case of the Weber problem with Euclidean distances; see page 97 in Chapter 6.

Since Algorithm 5.2 is computationally expensive unless additional in-formation is available on the structure of the problem, a heuristic strategy can alternatively be applied. In a large number of cases, this procedure still finds the optimal solution of $1/P/(B = N$ *convex polyhedra*$)/d_B/f$ *convex* and uses a remarkably smaller number of iterations. Instead of evaluating all the theoretically possible assignments of existing facilities to intermediate points, a suitable sample set S of points can be specified in $H_B \cap \mathcal{F}$. This can be done, for example, by choosing the grid points of an equidistant grid constructed in H_B, or by choosing specific points on the visibility grid $\mathcal{G}_d$. All the points $X \in S$ in this sample set are used to determine an objective function F_X according to Definition 5.3 and thus serve as starting points for an unconstrained location problem (5.8). As in Algorithm 5.2 the corresponding optimal solution Y^* is used as a candidate for the optimal solution of $1/P/(B = N\ convex\ polyhedra)/d_B/f\ convex$ if Y^* satisfies $f_B(Y^*) = F_X(Y^*)$.

Algorithm 5.3. (Approximation of $1/P/(B{=}N\ conv.\ pol.)/d_B/f\ conv.$**)**

Input: *Location problem* $1/P/(B = N\ convex\ polyhedra)/d_B/f\ convex$.

Step 1: *Construct the grid* $\mathcal{G}_d$.

Step 2: *Find the minima of* $1/P/(B = N\ convex\ polyhedra/d_B/ f\ convex\ on\ \mathcal{G}_d$.

Step 3: *Define a sample set* S *of points in* $H_B \cap \mathcal{F}$.

Step 4: *For each point* $X \in S$ *do:*

    (a) *Find an optimal solution* Y^* *of the corresponding unconstrained problem* $\min\{F_X(Y) : Y \in \mathbb{R}^2\}$.

    (b) *If* $f_B(Y^*) = F_X(Y^*)$, *the solution* Y^* *is a candidate for an optimal solution.*

Step 5: *Determine the best solution found in Steps 2 and 4.*

Output: *Approximation of the optimal solution of* $1/P/(B = N\ convex\ polyhedra)/d_B/f\ convex$.

As in Algorithm 5.2, the grid $\mathcal{G}_{l_2}$ can be used in Algorithm 5.3 instead of the grid $\mathcal{G}_d$ for any metric d induced by a norm.

Note that Algorithm 5.3 still uses the special structure of the problem as discussed in Section 5.1 even though it is based on an enumeration strategy. For each point in the sample set S the related unconstrained optimization problem is solved, thus allowing the detection of optimal solutions outside the selected sample set. Moreover, we can expect that an optimal solution is found by Algorithm 5.3 already for small sample sets S if these sets are selected appropriately. The detection of an optimal solution of a problem of type $1/P/(B = N\ convex\ polyhedra)/d_B/f\ convex$ can be guaranteed if, for example, a sufficiently fine sample set S of equidistant grid points is used:

Theorem 5.4. *For any problem of type* $1/P/(B = N\ convex\ polyhedra)/$ $d_B/f\ convex$, *Algorithm 5.3 finds an optimal solution if a sample set* S

of equidistant grid points with sufficiently small step length is chosen in $H_{\mathcal{B}} \cap \mathcal{F}$.

Proof. If an optimal solution of $1/P/(\mathcal{B} = N \text{ convex polyhedra})/d_{\mathcal{B}}$ f convex is located on the grid $\mathcal{G}_d$, then this solution is found in Step 2 of the algorithm.

Otherwise, consider a decomposition of $\mathcal{F}$ into a finite set of subregions of constant intermediate points $R_k \subseteq \mathcal{F}$, $k = 1, \ldots, K$, according to Definition 5.1, so that every subregion has a nonempty interior. Note that nonempty subregions with empty interior can be discarded, since they must be contained in the boundary of subregions with nonempty interior. Furthermore, let ε be maximal with the property that a ball of radius ε can be included in the interior of every subregion R_k, $k = 1, \ldots, K$. If a sample set S of equidistant grid points with step length ε is chosen, every subregion contains at least one grid point in its interior. Thus every feasible assignment of existing facilities to intermediate points leading to a reformulation of the objective function according to Definition 5.3 is considered at least once in the procedure. □

The sample set S can be chosen in many alternative ways. An intuitive option is to select sample points from the grid $\mathcal{G}_d$. However, a high solution quality implies a large number of iterations and thus a decreasing efficiency of the algorithm. A large sample set improves the quality of the solution, but on the other hand its size is proportional to the number of iterations of the procedure.

The performance of Algorithm 5.3 was compared for different sizes of the sample set S using an implementation based on LoLA (see Hamacher et al., 1999a; Ochs, 1998). The solutions were compared to an approximate global optimum obtained by evaluating the objective function at a finite set of equidistant points.

If, for example, only 10% of the intersection points in $\mathcal{G}_d$ are selected (according to their objective value, i.e., the best 10% of the intersection points are chosen), Algorithm 5.3 converged to an optimal solution in all of 54 sample problems with one circular barrier and from 20 to up to 60 existing facilities, whereas in the case of a thin rectangular barrier an optimal solution was found only in 25 out of 38 sample problems.

Summarizing the discussion above, if computation time is the major concern and if the special case of a Weber problem with the Euclidean metric $1/P/(\mathcal{B} = N \text{ convex polyhedra})/l_{2,\mathcal{B}}/\sum$ is considered, the iterative procedure developed in Butt and Cavalier (1996) is preferable, since it determines a solution that often is close to an optimal solution in a very small number of iterations. If, on the other hand, the quality of the solution is more important, or if a more general location problem is to be solved, Algorithm 5.3 (or Algorithm 5.2) can be applied, with an accuracy and computation time specified by the user with the choice of the sample set S.

It is an interesting open question whether it is possible to construct a small sample set S that still guarantees the detection of an optimal solution of the barrier problem $1/P/(\mathcal{B} = N\ convex\ polyhedra)/d_{\mathcal{B}}/f\ convex$.

5.4   Mixed-Integer Programming Formulations

The problem of assigning the optimal intermediate points I_m to the existing facilities Ex_m, $m \in \mathcal{M}$, with respect to an optimal solution in a given cell $C \in \mathcal{C}(\mathcal{G}_d)$ (or $C \in \mathcal{C}(\mathcal{G}_{l_2})$, respectively) can be formulated as a mixed-integer programming problem (MIP).

For a given cell $C \in \mathcal{C}(\mathcal{G}_d)$ with nonempty interior, let $\mathcal{I} \subseteq \mathcal{E}x \cup \mathcal{P}(\mathcal{B})$ be that subset of the candidates for intermediate points that are d-visible from all points in C, and without loss of generality let $\mathcal{I} = \{I_1, \ldots, I_k\}$ with $0 \leq k \leq |\mathcal{E}x| + |\mathcal{P}(\mathcal{B})|$. Furthermore, binary variables y_{im}, $i = 1, \ldots, k$, $m = 1, \ldots, M$, are defined as

$$y_{im} = \begin{cases} 1, & I_i \text{ is used as intermediate point } I_{Ex_m.X}, & i=1,\ldots,k, \\ 0, & I_i \text{ is not used as intermediate point } I_{Ex_m.X}, & m=1,\ldots,M. \end{cases}$$

Based on Theorem 5.1, problems of type $1/P/(\mathcal{B} = N\ convex\ polyhedra/d_{\mathcal{B}}/f\ convex$ can thus be represented by $|\mathcal{C}(\mathcal{G}_d)|$ mixed-integer programming formulations (5.11), each of them restricted to a cell $C \in \mathcal{C}(\mathcal{G}_d)$:

$$\min f\left(\sum_{i=1}^{k} y_{i1}[d(X,I_i)+d_{\mathcal{B}}(I_i, Ex_1)], \ldots, \sum_{i=1}^{k} y_{iM}[d(X,I_i)+d_{\mathcal{B}}(I_i, Ex_M)]\right)$$

$$\text{s.t.} \quad \sum_{i=1}^{k} y_{im} = 1 \qquad \forall m = 1, \ldots, M, \tag{5.11}$$

$$X \in C,$$

$$y_{im} \in \{0,1\} \qquad \forall i = 1, \ldots, k, \ m = 1, \ldots, M,$$

where the values of $d_{\mathcal{B}}(I_i, Ex_m)$ are known constants for all $i = 1, \ldots, k$ and $m = 1, \ldots, M$.

Even though the objective function of (5.11) is in general nonlinear, the constraint $X \in C$ of (5.11) can be formulated using only linear inequality constraints.

For this purpose recall that the grid $\mathcal{G}_{l_2}$ can be used to replace the grid $\mathcal{G}_d$ for any metric d induced by a norm (see Corollary 5.1). In this case, all cells in $\mathcal{C}(\mathcal{G}_{l_2})$ are convex, since the boundaries of all shadows are exclusively defined by half-lines starting in extreme points of the barriers and by facets of barriers. Moreover, each barrier facet is linearly extended by a half-line, since the extreme points of the facet also define a shadow region (cf. Figure 5.4 on page 62). Consequently, each cell $C \in \mathcal{C}(\mathcal{G}_{l_2})$ can be described by a finite number of linear inequality constraints. If, however, the grid $\mathcal{G}_d$ is used in the above formulation, we can still use the fact that all cells in $\mathcal{C}(\mathcal{G}_d)$ have piecewise linear boundaries; see Lemma 2.6 on page 35. If a cell $C \in \mathcal{C}(\mathcal{G}_d)$ is nonconvex, it can be decomposed into a finite series of convex subpolyhedra in each of which a mixed-integer programming problem (5.11) can be formulated with linear inequality constraints. For the decomposition of a nonconvex polyhedron into a finite (but not always minimal) set of convex subpolyhedra, efficient algorithms are given, for example, in Fernández et al. (1997, 2000).

In the case of a Weber objective function, the MIP problem (5.11) simplifies to

$$\min \quad \sum_{m=1}^{M} \left(\sum_{i=1}^{k} y_{im} w_m [d(X, I_i) + d_{\mathcal{B}}(I_i, Ex_m)] \right)$$

$$\text{s.t.} \quad \sum_{i=1}^{k} y_{im} = 1 \qquad \forall m = 1, \ldots, M, \qquad (5.12)$$

$$X \in C,$$

$$y_{im} \in \{0, 1\} \qquad \forall i = 1, \ldots, k, \ m = 1, \ldots, M,$$

which can be formulated with a quadratic objective function in the case that the prescribed metric is a block norm γ with the fundamental vectors $v^1, \ldots, v^{\delta}$ (cf. Lemma 1.4 on page 8):

$$\min \quad \sum_{m=1}^{M} \left(\sum_{i=1}^{k} y_{im} w_m \left[\sum_{j=1}^{\delta} \lambda_{ji} + \gamma_{\mathcal{B}}(I_i, Ex_m) \right] \right)$$

$$\text{s.t.} \quad \sum_{j=1}^{\delta} \lambda_{ji} \, v^j = X - I_i \qquad \forall i = 1, \ldots, k,$$

$$\sum_{i=1}^{k} y_{im} = 1 \qquad \forall m = 1, \ldots, M, \qquad (5.13)$$

$$X \in C,$$

$$y_{im} \in \{0, 1\} \qquad \forall i = 1, \ldots, k, \ m = 1, \ldots, M,$$

$$\lambda_{ji} \geq 0 \qquad \forall i = 1, \ldots, k, \ j = 1, \ldots, \delta.$$

For the center objective function, (5.11) can be written as

$$
\min \quad \max_{m=1,\dots,M} \left(\sum_{i=1}^{k} y_{im} w_m [d(X,I_i)+d_\mathcal{B}(I_i,Ex_m)] \right)
$$

$$
\text{s.t.} \quad \sum_{i=1}^{k} y_{im} = 1 \qquad\qquad \forall m=1,\dots,M. \tag{5.14}
$$

$$
X \in C,
$$

$$
y_{im} \in \{0,1\} \qquad\qquad \forall i=1,\dots,k,\ m=1,\dots,M.
$$

Eliminating the maximization in the objective function, we obtain the following equivalent representation of (5.14):

$$
\min \ z
$$

$$
\text{s.t.} \quad z \geq \sum_{i=1}^{k} y_{im} w_m [d(X,I_i)+d_\mathcal{B}(I_i,Ex_m)] \quad \forall m=1,\dots,M.
$$

$$
\sum_{i=1}^{k} y_{im} = 1 \qquad\qquad \forall m=1,\dots,M, \tag{5.15}
$$

$$
X \in C,
$$

$$
y_{im} \in \{0,1\} \qquad\qquad \forall i=1,\dots,k,\ m=1,\dots,M,
$$

$$
z \in \mathbb{R}_+.
$$

Note that in this formulation the constraints $z \geq \sum_{i=1}^{k} y_{im} w_m [d(X,I_i)+ d_\mathcal{B}(I_i,Ex_m)]$, $m=1,\dots,M$, are in general nonlinear, depending on the choice of d.

If a mixed-integer programming problem (5.11) is solved for each cell $C \in \mathcal{C}(\mathcal{G}_d)$, the set of optimal solutions of problems of type $1/P/(\mathcal{B} = N$ convex polyhedra$)/d_\mathcal{B}/f$ convex can be determined from the individual minimizers by the following algorithm:

Algorithm 5.4. (MIP for $1/P/(\mathcal{B} = N$ conv. pol.$)/d_\mathcal{B}/f$ conv.)

Input: *Location problem $1/P/(\mathcal{B} = N$ convex polyhedra$)/d_\mathcal{B}/f$ convex.*

Step 1:   *Construct the grid $\mathcal{G}_d$.*

Step 2:   *For each cell $C \in \mathcal{C}(\mathcal{G}_d)$ do:*
   (a) *Find that subset $\mathcal{I} \subseteq \mathcal{E}x \cup P(\mathcal{B})$ of intermediate points that are d-visible from all points in C.*

(b) *Find an optimal solution of the MIP formulation (5.11) in the cell C.*

Step 3: *Determine the best solution found in Step 2.*

Output: *Set of optimal solutions of $1/P/(\mathcal{B} = N \text{ conv. pol.})/d_\mathcal{B}/f \text{ conv.}$*

The correctness of Algorithm 5.4 follows from Theorem 5.1 and from the above MIP formulations. As in Algorithms 5.2 and 5.3, the grid $\mathcal{G}_d$ can be replaced by the grid $\mathcal{G}_{l_2}$.

The time complexity of Algorithm 5.4 depends on the number of cells in $\mathcal{C}(\mathcal{G}_d)$ and on the complexity of solving the mixed-integer programming problems (5.11) in each cell, which in turn depends on the objective function f and the choice of the distance function d. If $d = l_2$, the number of cells in $\mathcal{G}_{l_2}$ is bounded by $O\left((|\mathcal{E}x| + |\mathcal{P}(\mathcal{B})|)^2 \cdot |\mathcal{P}(\mathcal{B})|^2\right)$; see also the discussion of the grid $\mathcal{G}_{l_2}$ on page 61 and the discussion of Algorithm 5.2 on page 71.

5.5 Weber Objective Functions

In the special case of the Weber objective function

$$f_\mathcal{B}(X) = \sum_{m=1}^{M} w_m d_\mathcal{B}(X, Ex_m)$$

with nonnegative weights $w_m \in \mathbb{R}_+$, $m = 1, \ldots, M$, the results of Section 5.1 can be further simplified. Even though we will formulate the following results only for the Weber objective function, they can be easily generalized to all objective functions f that are homomorphisms satisfying

$$f(d(X, I_1) + d(I_1, Ex_1), \ldots, d(X, I_M) + d(I_M, Ex_M))$$
$$= f(d(X, I_1), \ldots, d(X, I_M)) + f(d(I_1, Ex_1), \ldots, d(I_M, Ex_M)) \tag{5.16}$$

for every choice of $I_1, \ldots, I_M$ and X in $\mathbb{R}^2$.

In this case, the objective function can be separated into one part depending on the distances from X to the intermediate points and into a constant part based on the distances from the intermediate points to the existing facilities.

Corollary 5.3. *Let $C \in \mathcal{C}(\mathcal{G}_d)$ be a cell and let $X \in C$ be a feasible solution of $1/P/(\mathcal{B} = N \text{ convex polyhedra})/d_\mathcal{B}/\sum$. Then*

$$f_\mathcal{B}(X) = f_X(X) + g_X, \tag{5.17}$$

where

$$f_X(Y) \ := \ \sum_{m=1}^{M} w_m d(Y, I_m), \qquad Y \in \mathbb{R}^2, \qquad (5.18)$$

$$g_X \ := \ \sum_{m=1}^{M} w_m d_{\mathcal{B}}(I_m, Ex_m), \qquad\qquad (5.19)$$

and $I_m := I_{Ex_m, X}$, $i = 1, \ldots, M$, is an intermediate point on a d-shortest permitted Ex_m-X path according to Definition 2.5 that is d-visible from all points in C.

Note that g_X can alternatively be determined as $g_X = \sum_{m=1}^{M} w_m d_G(I_m, Ex_m)$, using the network distance $d_G(I_m, Ex_m)$ between I_m and Ex_m in the visibility graph of $\mathcal{E}x \cup \mathcal{P}(\mathcal{B})$.

Again, $f_X(Y)$ is a convex function in $\mathbb{R}^2$, since it does not depend on the nonconvex distance function $d_{\mathcal{B}}$. Furthermore, g_X is a constant not depending on the choice of Y. Therefore, $f_X(Y) + g_X$ is also a convex function of Y in $\mathbb{R}^2$.

Like Theorems 5.1 and 5.2, Corollary 5.3 can be used to interrelate $1/P/(\mathcal{B} = N$ *convex polyhedra*$)/d_{\mathcal{B}}/ \sum$ to a finite set of the corresponding unconstrained problems $1/P/ \bullet /d/ \sum$.

Theorem 5.5. *Let* $C \in \mathcal{C}(\mathcal{G}_d)$ *be a cell and let* $X_{\mathcal{B}}^* \in C$ *be an optimal solution of* $1/P/(\mathcal{B} = N$ *convex polyhedra*$)/d_{\mathcal{B}}/ \sum$. *Then* $X_{\mathcal{B}}^*$ *is an optimal solution of the corresponding restricted location problem*

$$\begin{aligned} \min \quad & f_{X_{\mathcal{B}}^*}(Y) + g_{X_{\mathcal{B}}^*} \\ \text{s.t.} \quad & Y \in C, \end{aligned} \qquad (5.20)$$

where $f_{X_{\mathcal{B}}^*}(Y)$ *and* $g_{X_{\mathcal{B}}^*}$ *are defined according to (5.18) and (5.19), respectively.*

Moreover, if $X_{\mathcal{B}}^* \notin \mathcal{G}_{l_2}$ *is an optimal solution of* $1/P/(\mathcal{B} = N$ *convex polyhedra*$)/d_{\mathcal{B}}/ \sum$, *then* $X_{\mathcal{B}}^*$ *is an optimal solution of the corresponding unconstrained convex problem*

$$\begin{aligned} \min \quad & f_{X_{\mathcal{B}}^*}(Y) + g_{X_{\mathcal{B}}^*} \\ \text{s.t.} \quad & Y \in \mathbb{R}^2. \end{aligned} \qquad (5.21)$$

Proof. Let $C \in \mathcal{C}(\mathcal{G}_d)$, $X_{\mathcal{B}}^* \in C$, and let $f_{X_{\mathcal{B}}^*}(Y)$ and $g_{X_{\mathcal{B}}^*}$ be defined according to (5.18) and (5.19). Furthermore let I_m, $m \in \mathcal{M}$, be the corresponding intermediate points on d-shortest permitted Ex_m-$X_{\mathcal{B}}^*$ paths that are d-visible from all points in C. Using (5.3) we obtain that

$$F_{X_{\mathcal{B}}^*}(Y) = \sum_{i=1}^{M} w_m (d(Y, I_m) + d_{\mathcal{B}}(I_m, Ex_m)) = f_{X_{\mathcal{B}}^*}(Y) + g_{X_{\mathcal{B}}^*},$$

and thus Theorems 5.1 and 5.2 imply the result.   □

Similarly to Corollary 5.1, Theorem 5.5 can be reformulated with respect to the grid $\mathcal{G}_{l_2}$:

Corollary 5.4. *Let d be a distance function induced by a norm. Furthermore, let $C \in \mathcal{C}(\mathcal{G}_{l_2})$ be a cell in the grid $\mathcal{G}_{l_2}$ and let $X_{\mathcal{B}}^* \in C$ be an optimal solution of $1/P/(\mathcal{B} = N$ convex polyhedra$)/d_{\mathcal{B}}/\sum$. Then $X_{\mathcal{B}}^*$ is an optimal solution of the corresponding restricted problem*

$$
\begin{aligned}
\min \quad & f_{X_{\mathcal{B}}^*}(Y) + g_{X_{\mathcal{B}}^*} \\
\text{s.t.} \quad & Y \in C,
\end{aligned}
\tag{5.22}
$$

where $f_{X_{\mathcal{B}}^*}(Y)$ and $g_{X_{\mathcal{B}}^*}$ are defined according to (5.18) and (5.19) and the intermediate points I_m ($m \in \mathcal{M}$) are chosen such that they additionally are l_2-visible from $X_{\mathcal{B}}^*$.

If $X_{\mathcal{B}}^* \notin \mathcal{G}_{l_2}$ is an optimal solution of $1/P/(\mathcal{B} = N$ convex polyhedra$)/d_{\mathcal{B}}/\sum$, then $X_{\mathcal{B}}^*$ is an optimal solution of the corresponding unconstrained, convex problem

$$
\begin{aligned}
\min \quad & f_{X_{\mathcal{B}}^*}(Y) + g_{X_{\mathcal{B}}^*} \\
\text{s.t.} \quad & Y \in \mathbb{R}^2.
\end{aligned}
\tag{5.23}
$$

5.6 Multifacility Weber Problems

This section gives an outline of how the results derived up to now can be generalized to problems of much higher complexity without the need for an entirely new theoretical approach.

In a multifacility Weber problem the optimal location of K new facilities $(X_1, \ldots, X_K)$ is sought with respect to a finite set of existing facilities $\mathcal{E}x = \{Ex_1, \ldots, Ex_M\}$ in the feasible region $\mathcal{F} \subseteq \mathbb{R}^2$. The interaction between the existing facilities and the new facilities and the interaction between pairs of new facilities can be modeled by a connected graph (F, E) with node set F and edge set E. The node set F can be partitioned into two sets A, V such that $F = A \cup V$ and $A \cap V = \emptyset$, where each node $a \in A$ represents an existing facility $X_a \in \mathcal{E}x$, and each node $v \in V$ corresponds to a new facility $X_v \in \{X_1, \ldots, X_K\}$. Furthermore, a positive weight w_e is associated with each edge $e = \{u_1, u_2\} \in E$ representing the interaction between two facilities $u_1 \in V$ and $u_2 \in F$. Then the *multifacility Weber problem with barriers*, classified as $K/P/(\mathcal{B} = N$ convex polyhedra$)/d_{\mathcal{B}}/\sum$, can be formulated as

$$
\begin{aligned}
\min \quad & f_B(X_1, \ldots, X_K) = \sum_{e=\{u_1, u_2\} \in E} w_e \, d_{\mathcal{B}}(X_{u_1}, X_{u_2}) \\
\text{s.t.} \quad & X_1, \ldots, X_K \in \mathcal{F}.
\end{aligned}
\tag{5.24}
$$

In the following we will generalize the results of Sections 5.1 and 5.5 to multifacility Weber problems with barriers.

Lemma 5.7. Let $\underline{X} = (X_1, \ldots, X_K)$ be a feasible solution of the multifacility Weber problem with barriers (5.24). Then the objective function f_B can be rewritten as

$$f_B(\underline{X}) = f_{\underline{X}}(\underline{X}) + g_{\underline{X}}, \tag{5.25}$$

where $f_{\underline{X}}(\underline{Y})$ is the convex objective function of an unconstrained multifacility Weber problem and $g_{\underline{X}}$ is a constant not explicitly depending on $\underline{X}$.

Proof. Let $\underline{X} = (X_1, \ldots, X_K)$ be a feasible solution of (5.24). For simplicity we denote for two points $X_u, X_v \in \mathcal{F}$ an intermediate point I_{X_u, X_v} by $I_{u,v}$ and I_{X_v, X_u} by $I_{v,u}$.

Using Lemma 2.3 on page 31 and the reformulation of the barrier distance according to Corollary 2.3 on page 34 we can conclude that there exist intermediate points according to Definition 2.5 such that the objective function of (5.24) can be rewritten as

$$
\begin{aligned}
f_B(\underline{X}) \;=\; & \sum_{e=\{u_1,u_2\}\in E} w_e d_B(X_{u_1}, X_{u_2}) \\[2mm]
=\; & \sum_{\substack{e=\{v,a\}\in E:\\ v\in V,\, a\in A}} w_e d_B(X_v, X_a) + \sum_{\substack{e=\{u,v\}\in E:\\ u,v\in V}} w_e d_B(X_u, X_v) \\[2mm]
=\; & \sum_{\substack{e=\{v,a\}\in E:\\ v\in V,\, a\in A}} w_e \big(d_G(X_a, I_{a,v}) + d(I_{a,v}, X_v)\big) \\[2mm]
& + \sum_{\substack{e=\{u,v\}\in E:\ u,v\in V,\\ X_v\in \text{visible}_d(X_u)}} w_e d(X_u, X_v) \\[2mm]
& + \sum_{\substack{e=\{u,v\}\in E:\ u,v\in V,\\ X_v\in \text{shadow}_d(X_u)}} w_e \big(d(X_u, I_{v,u}) + d_G(I_{v,u}, I_{u,v}) + d(I_{u,v}, X_v)\big).
\end{aligned}
$$

It can be easily verified that the definition of

$$
\begin{aligned}
f_{\underline{X}}(\underline{X}) \;:=\; & \sum_{\substack{e=\{v,a\}\in E:\\ v\in V,\, a\in A}} w_e d(I_{a,v}, X_v) + \sum_{\substack{e=\{u,v\}\in E:\ u,v\in V,\\ X_v\in \text{visible}_d(X_u)}} w_e d(X_u, X_v) \\[2mm]
& + \sum_{\substack{e=\{u,v\}\in E:\ u,v\in V,\\ X_v\in \text{shadow}_d(X_u)}} w_e \big(d(X_u, I_{v,u}) + d(I_{u,v}, X_v)\big)
\end{aligned} \tag{5.26}
$$

and

$$g_{\underline{X}} := \sum_{\substack{e=\{v,a\}\in E:\\ v\in V,\, a\in A}} w_e d_G(X_a, I_{a,v}) + \sum_{\substack{e=\{u,v\}\in E:\, u,v\in V,\\ X_v\in \text{shadow}_d(X_u)}} w_e d_G(I_{v,u}, I_{u,v})$$

(5.27)

yields a convex function $f_{\underline{X}}(\underline{Y})$ that does not depend on $d_{\mathcal{B}}$. In particular, $f_{\underline{X}}(\underline{Y})$ can be interpreted as the objective function of an unconstrained multifacility Weber problem with existing facilities at the points $Ex_1, \ldots, Ex_M$, $I_{a,v}$ for $e = \{v, a\} \in E : v \in V, a \in A$, and $I_{u,v}, I_{v,u}$ for $e = \{u, v\} \in E : u, v \in V, X_v \in \text{shadow}_d(X_u)$. Moreover, $g_{\underline{X}}$ can be interpreted as a constant not explicitly depending on $\underline{X}$. □

Theorem 5.6. *Let $\underline{X}_{\mathcal{B}}^* = (X_1^*, \ldots, X_K^*)$ be an optimal solution of problem (5.24). Then $\underline{X}_{\mathcal{B}}^*$ is an optimal solution of the corresponding unconstrained multifacility Weber problem*

$$\begin{aligned} \min \quad & f_{\underline{X}_{\mathcal{B}}^*}(Y_1, \ldots, Y_N) + g_{\underline{X}_{\mathcal{B}}^*}\\ \text{s.t.} \quad & Y_1, \ldots, Y_K \in \mathbb{R}^2, \end{aligned}$$

(5.28)

where $f_{\underline{X}_{\mathcal{B}}^*}(Y_1, \ldots, Y_K)$ and $g_{\underline{X}_{\mathcal{B}}^*}$ are defined according to (5.26) and (5.27), respectively, if

$$X_1^*, \ldots, X_K^* \notin \left(\mathcal{G}_d \cup \bigcup_{i=1,\ldots,K} \partial(\text{shadow}_d(X_i^*)) \right).$$

(5.29)

Proof. Let $\underline{X}_{\mathcal{B}}^* = (X_1^*, \ldots, X_K^*)$ be an optimal solution of (5.24) satisfying (5.29). For $i = 1, \ldots, K$ there exists an $\varepsilon_i > 0$ such that the ε_i-neighborhood $N_{\varepsilon_i}(X_i^*) = \{Z \in \mathbb{R}^2 : l_2(Z, X_i^*) < \varepsilon_i\}$ of X_i^*, $i = 1, \ldots, K$, is not intersected by a boundary of a shadow of either an existing facility, an extreme point of a barrier, or another new location $X_j^*, j \in \{1, \ldots, K\}$, $j \neq i$. Thus all candidates for intermediate points on d-shortest permitted paths to existing facilities or to other new locations that are d-visible from X_i^* are also d-visible from all points in $N_{\varepsilon_i}(X_i^*)$, $i = 1, \ldots, K$. Therefore, there exists $\bar{\varepsilon} \leq \min\{\varepsilon_1, \ldots, \varepsilon_K\}$, $\bar{\varepsilon} > 0$, such that the following inequality holds for all $\underline{Y} = (Y_1, \ldots, Y_K)$ satisfying $Y_i \in N_{\bar{\varepsilon}}(X_i^*)$ for all $i = 1, \ldots, K$ (cf. Lemma 5.5):

$$f_{\mathcal{B}}(\underline{Y}) = f_{\underline{Y}}(\underline{Y}) + g_{\underline{Y}} \leq f_{\underline{X}_{\mathcal{B}}^*}(\underline{Y}) + g_{\underline{X}_{\mathcal{B}}^*}.$$

(5.30)

Assume that there exists a solution $\underline{Y}^* = (Y_1^*, \ldots, Y_K^*)$ with $Y_i^* \in N_{\bar{\varepsilon}}(X_i^*)$, $i = 1, \ldots, K$, such that

$$f_{\underline{X}_{\mathcal{B}}^*}(\underline{Y}^*) + g_{\underline{X}_{\mathcal{B}}^*} < f_{\underline{X}_{\mathcal{B}}^*}(\underline{X}_{\mathcal{B}}^*) + g_{\underline{X}_{\mathcal{B}}^*}.$$

Using (5.30) we can conclude that

$$f_B(\underline{Y}^*) = f_{\underline{Y}^*}(\underline{Y}^*) + g_{\underline{Y}^*} \leq f_{\underline{X}_B^*}(\underline{Y}^*) + g_{\underline{X}_B^*}$$
$$< f_{\underline{X}_B^*}(\underline{X}_B^*) + g_{\underline{X}_B^*} = f_B(\underline{X}_B^*),$$

contradicting the optimality of $\underline{X}_B^*$. Since $f_{\underline{X}_B^*}(\underline{Y})$ is convex and $g_{\underline{X}_B^*}$ is constant (cf. Lemma 5.7), it follows that $\underline{X}_B^*$ minimizes $f_{\underline{X}_B^*}(\underline{Y}) + g_{\underline{X}_B^*}$ in $\mathbb{R}^2$.   □

Note that the restriction of Theorem 5.6 to those solutions for which the new facilities are not located on the boundary of a shadow of any existing facility, any extreme point of a barrier, or any other new facility location (5.29) is only partially used in the proof of Theorem 5.6. A weaker condition for the applicability of Theorem 5.6 is the assumption that every new facility $X_v \in \{X_1, \ldots, X_K\}$ is not located on the boundary of the shadow of any other point $X_u \in \mathcal{E}x \cup \mathcal{P}(\mathcal{B})$ with which it interacts with respect to the modified objective function $f_{\underline{X}}$.

Alternatively, condition (5.29) can be replaced by the following condition (5.31):

$$X_1^*, \ldots, X_K^* \notin \left(\mathcal{G}_{l_2} \cup \bigcup_{i=1,\ldots,K} \partial(\text{shadow}_{l_2}(X_i^*)) \right). \tag{5.31}$$

In this case intermediate points on d-shortest permitted X-Y paths have to be chosen such that they are l_2-visible from Y in the proof of Theorem 5.6.

Theorem 5.6 implies that if an optimal reformulation of the objective function with respect to (5.26) and (5.27) is known, i.e., if an optimal assignment of interactions between new and existing facilities to intermediate points is known, the nonconvex multifacility Weber problem with barriers (5.24) can be solved by means of one related unconstrained and thus convex multifacility Weber problem (5.28).

The corresponding unconstrained multifacility Weber problems $K/P/ \bullet /d/\sum$ can be solved, for example, by the polynomial-time algorithm developed in Fliege (2000). For further results on multifacility Weber problems, see, for example, Fliege (1997) and Plastria (1995).

5.7   Related Problems

The results developed in the previous sections can be easily adapted to the case that, in addition to the barrier sets, a finite set of forbidden regions $\{R_1, \ldots, R_L\}$ is given in which the location of a new facility is forbidden, while traveling through these areas is allowed. In this case the subproblems resulting from the reformulation of the objective function in Section 5.1 are restricted location problems with the forbidden regions $R_1, \ldots, R_L$.

Another generalization that is of increasing interest in many practical applications is the introduction of more than one decision-maker and thus of multiple ($Q \geq 2$) objective functions. In this case the set of optimal solutions depends on the choice of the ordering of $\mathbb{R}^Q$ that is used to compare all feasible solutions. Its determination is easy if this ordering is given by the max-ordering or by the lexicographic ordering. If the solution vectors are compared componentwise, i.e., if the concept of Pareto optimality is used, further investigations are needed. Problems of this type will be discussed in detail in Chapter 10.

Three-dimensional location problems with barriers may occur, for example, in the optimal positioning of radio stations, or in the aerospace industry. A generalization to higher-dimensional problems seems to be more of theoretical than of practical interest. Anyhow, a generalization is possible also in this case if all the d-shortest permitted paths in $\mathbb{R}^n$, $n > 2$, are approximated by piecewise linear paths with breaking points only in a specified finite set of candidate points on the faces of the barrier polyhedra. This discretization approach is frequently used in the literature in the context of shortest-path problems in $\mathbb{R}^n$ with polyhedral barriers (compare the discussion on page 34 in Section 2.4), and it has its justification in Lemma 2.4; see page 31.

6

Location Problems with a Circular Barrier

Up to now the advantages of polyhedral barriers have been exploited by using the extreme points of the barrier sets as reference points for related unconstrained location problems. In the case of nonpolyhedral barriers, a different approach will be needed.

Based on the work of Katz and Cooper (1981), in this chapter we concentrate on Weber problems with one circular barrier and assume that distances are measured by the Euclidean metric. Even though we restrict ourselves to the case that only one barrier is given in the plane $\mathbb{R}^2$, some of the results of this chapter are more general and can also be transferred to the case that more than one barrier is given. Moreover, similar results can be expected if the given metric is an l_p metric with $p \in (1, \infty)$.

Let one circular barrier

$$\mathcal{B} = B_{C(r)} := \{X \in \mathbb{R}^2 : \|X\|_{l_2} \leq r\}$$

be given in the plane $\mathbb{R}^2$ that is, without loss of generality, centered at the origin and that has a finite positive radius $r \in \mathbb{R}_+$. As usual, $B_{C(r)}$ represents that region in the plane where neither trespassing nor the location of new facilities is allowed. Without loss of generality we can assume that the radius r of the circular barrier is equal to 1 and refer to $B_{C(r)}$ as $B_C := B_{C(1)}$. The feasible region $\mathcal{F}$ for new locations is given by $\mathcal{F} = \mathbb{R}^2 \setminus \text{int}(B_C)$. Furthermore, a finite set of existing facilities $\mathcal{E}x = \{Ex_m \in \mathcal{F} : m \in \mathcal{M}\}$ with the index set $\mathcal{M} = \{1, \ldots, M\}$ is given in the feasible region $\mathcal{F}$.

If distances are measured by the Euclidean metric l_2, and if the corresponding barrier distance l_{2,B_C} is defined according to Definition 2.2, a lo-

cation problem with a circular barrier $1/P/(\mathcal{B} = 1\,circle)/l_{2,B_C}/f\,convex$, or $1/P/B_C/l_{2,B_C}/f\,convex$ for short, can be formulated as

$$\min \quad f_B(X) = f(l_{2,B_C}(X, Ex_1), \ldots, l_{2,B_C}(X, Ex_M))$$
$$\text{s.t.} \quad X \in \mathcal{F}. \tag{6.1}$$

As in Chapter 5, we assume that the objective function $f(X) = f(l_2(X, Ex_1), \ldots, l_2(X, Ex_M))$ of the corresponding unconstrained problem is a convex and nondecreasing function of the distances $l_2(X, Ex_1), \ldots, l_2(X, Ex_M)$.

Some properties of l_2-shortest permitted paths in the presence of a circular barrier are derived in the following section. Furthermore, the convexity of the objective function f_B on certain open convex subsets of $\mathcal{F}$ is proven. Algorithmic consequences of these results are discussed in Section 6.2.

6.1   Properties of the Objective Function

The aim of this section is to identify subsets of the feasible region $\mathcal{F}$ on which convexity of the objective function f_B of the generally nonconvex problem (6.1) can be proven. As a first step towards this goal, we will discuss some properties of the barrier distance $l_{2,B_C}(X, Ex)$ between an arbitrary but fixed existing facility $Ex \in \mathcal{E}x$ and a point $X \in \mathcal{F}$.

First assume that $X \notin \text{shadow}_{l_2}(Ex)$; i.e., the point X is l_2-visible from the existing facility Ex.

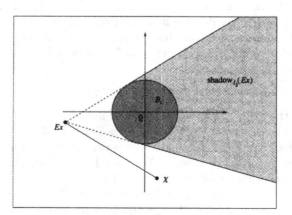

Figure 6.1. The l_2-shadow of an existing facility $Ex \in \mathcal{E}x$.

But then, the straight line segment connecting X and Ex is a permitted X-Ex path, and therefore $l_{2,B_C}(X, Ex) = l_2(X, Ex)$ holds in this case; see Figure 6.1.

If, on the other hand, $X \in \text{shadow}_{l_2}(Ex)$, then Theorem 2.1 and Corollary 2.2 of Section 2.2 apply for the calculation of $l_{2,B_C}(X, Ex)$.

According to Corollary 2.2 an l_2-shortest permitted X-Ex path consists of straight line segments and of circular sections on the boundary $\partial(B_C)$ of the barrier. Moreover, it was proven in Elsgolc (1962) (see also Katz and Cooper, 1981) that the straight line segments of an optimal path must be tangent to the boundary $\partial(B_C)$ of the circular barrier at every point of intersection. Since only one circular barrier is given in our case, we obtain two candidates for an optimal X-Ex path, as illustrated in Figure 6.2.

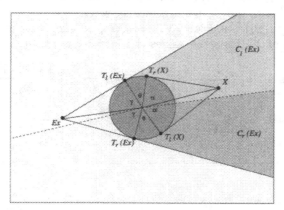

Figure 6.2. Two candidates for an l_2-shortest permitted X-Ex path.

In the following we will refer to the *right point of tangency* with respect to a given point $X \in \mathcal{F}$ as $T_r(X)$ and to the *left point of tangency* as $T_l(X)$, respectively, where "right" and "left" are defined with respect to the half-line starting at X and passing through the center of B_C, which in our case is the origin.

Obviously, the path through the points of tangency $T_l(Ex)$ and $T_r(X)$ is optimal in the example given in Figure 6.2.

The length of the l_2-shortest permitted X-Ex path in Figure 6.2 can be calculated as (see Katz and Cooper, 1981)

$$
\begin{aligned}
l_{2,B_C}(X, Ex) &= l_2(Ex, T_l(Ex)) + 2r \arcsin\left(\frac{l_2(T_l(Ex), T_r(X))}{2r}\right) \\
&\quad + l_2(T_r(X), X) \\[2mm]
&= l_2(Ex, T_l(Ex)) + r\theta + l_2(T_r(X), X),
\end{aligned}
\tag{6.2}
$$

where θ is the angle (in radians) enclosed by the two line segments between the origin and the two points of tangency $T_l(Ex)$ and $T_r(X)$.

Lemma 6.1. *Let $Ex \in \mathcal{E}x$ be an existing facility and let $X \in \mathcal{F}$. Then*

$$l_{2,B_C}(X, Ex) = \begin{cases} l_2(X, Ex) & \text{if } X \notin \text{shadow}_{l_2}(Ex), \\ l_2(Ex, T(Ex)) + r\theta + l_2(T(X), X) & \text{if } X \in \text{shadow}_{l_2}(Ex), \end{cases}$$

where $(T(Ex), T(X)) \in \{(T_r(Ex), T_l(X)), (T_l(Ex), T_r(X))\}$ such that the value of $l_{2,B_C}(X, Ex)$ is minimal.

In the example depicted in Figure 6.3, the points of tangency $T_r(X) = (t_1^r, t_2^r)^T$ and $T_l(X) = (t_1^l, t_2^l)^T$ for a given point $X \in \mathcal{F}$ can be obtained, using the angles α and β, in the following way:

Let $X = (x_1, x_2)^T$ and let $\|X\| := \|X\|_{l_2} = \sqrt{x_1^2 + x_2^2}$ denote the Euclidean norm of X. Then the angles α and β are given by

$$\cos\alpha = \frac{r}{\|X\|}, \qquad \sin\alpha = \frac{\sqrt{\|X\|^2 - r^2}}{\|X\|},$$

$$\cos\beta = \frac{x_1}{\|X\|}, \qquad \sin\beta = \frac{x_2}{\|X\|}.$$

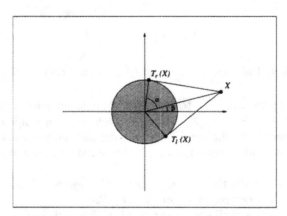

Figure 6.3. Determination of the points of tangency $T_r(X)$ and $T_l(X)$.

Depending on whether we want to find the right point of tangency $T_r(X)$ or the left point of tangency $T_l(X)$, the values of the coordinates t_1 and t_2 of the respective point of tangency are given by (see Katz and Cooper, 1981)

$$t_1^{r,l} = r\cos(\beta \pm \alpha) = r(\cos\alpha\cos\beta \mp \sin\alpha\sin\beta),$$

$$t_2^{r,l} = r\sin(\beta \pm \alpha) = r(\cos\alpha\sin\beta \pm \sin\alpha\cos\beta).$$

Summarizing the discussion above, we obtain the following representation of the right and left points of tangency, respectively:

$$T_r(X) = \left(\frac{r^2 x_1 - r x_2 \sqrt{x_1^2 + x_2^2 - r^2}}{x_1^2 + x_2^2}, \quad \frac{r^2 x_2 + r x_1 \sqrt{x_1^2 + x_2^2 - r^2}}{x_1^2 + x_2^2} \right)^T,$$

$$\tag{6.3}$$

$$T_l(X) = \left(\frac{r^2 x_1 + r x_2 \sqrt{x_1^2 + x_2^2 - r^2}}{x_1^2 + x_2^2}, \quad \frac{r^2 x_2 - r x_1 \sqrt{x_1^2 + x_2^2 - r^2}}{x_1^2 + x_2^2} \right)^T.$$

Given a point $X \in \mathrm{shadow}_{l_2}(Ex)$ as depicted in Figure 6.2, an l_2-shortest permitted X-Ex path passes through the points of tangency $T_l(Ex), T_r(X)$ or through $T_r(Ex), T_l(X)$, depending on the relative magnitude of the angles θ and ϕ.

For every existing facility $Ex \in \mathcal{E}x$, let a construction line $(\underline{0} - Ex)_{B_C}$ be defined as

$$(\underline{0} - Ex)_{B_C} := \{-\lambda Ex \ : \ \lambda \in \mathbb{R}_+; \ \lambda \|Ex\| \geq r\}. \tag{6.4}$$

Then the construction line $(\underline{0} - Ex)_{B_C}$ decomposes the set $\mathrm{shadow}_{l_2}(Ex)$ into two cells $C_r(Ex)$ and $C_l(Ex)$ (see Figure 6.2) such that for every point $X \in C_r(Ex)$ an l_2-shortest permitted X-Ex path passes through $T_r(Ex), T_l(X)$, and for every point $X \in C_l(Ex)$ an l_2-shortest permitted X-Ex path passes through $T_l(Ex), T_r(X)$. Observe that for all points on the construction line $(\underline{0} - Ex)_{B_C}$ both paths have the same length; i.e., an l_2-shortest permitted X-Ex path may pass along either side of the barrier.

Even though the points $T_r(Ex)$ and $T_l(Ex)$ depend only on the fixed location of the existing facility $Ex \in \mathcal{E}x$ and are therefore also fixed, the locations of the points of tangency $T_r(X)$ and $T_l(X)$ depend on the coordinates of X and change when X is moved in the set $\mathrm{shadow}_{l_2}(Ex)$. This behavior complicates the computation of the barrier distance in the case of round barrier sets compared to polyhedral barrier sets, where constant intermediate points could be used to describe the barrier distance (cf. Section 2.4). However, the following result proves that the barrier distance $l_{2,B_C}(Ex, X)$ is nevertheless convex on every open convex subset of a cell $C_r(Ex)$ or $C_l(Ex)$, respectively.

Lemma 6.2. *Let $Ex \in \mathcal{E}x$ be an existing facility and let $C_r(Ex)$ and $C_l(Ex)$ be the two cells obtained from subdividing the set $\mathrm{shadow}_{l_2}(Ex)$ with the construction line $(\underline{0} - Ex)_{B_C}$. Then $l_{2,B_C}(X, Ex)$ is a convex function of X on every open convex set O satisfying $O \subset C_r(Ex)$ or $O \subset C_l(Ex)$.*

Proof. To simplify further notation we will use in the following that without loss of generality $r = 1$, and assume that the existing facility $Ex \in \mathcal{E}x$ is located in the fourth quadrant of the Cartesian coordinate system such that $Ex = (a_1, a_2)^T$ with $a_1 = 1$ and $a_2 < 0$. Thus the right point of tangency $T_r(Ex)$ to the circle is given by $T_r(Ex) = (1, 0)^T$; see Figure 6.4. Due to the symmetry of the problem it is sufficient to prove the convexity of $l_{2,B_C}(X, Ex)$ on every open convex subset O of the cell $C_r(Ex)$.

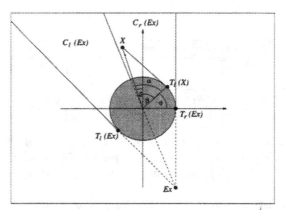

Figure 6.4. The special situation considered in the proof of Lemma 6.2.

Under these assumptions, the distance $l_{2,B_C}(X,Ex)$ between an arbitrary point $X = (x_1, x_2)^T \in C_r(Ex)$ and the existing facility $Ex = (1, a_2)^T$ can be calculated as

$$
\begin{aligned}
l_{2,B_C}(X,Ex) &= l_2(X,T_l(X)) + r\phi + l_2(T_r(Ex),Ex) \\
&= g_1(X) + g_2(X) + |a_2| \\
&=: g(X),
\end{aligned}
$$

where $l_2(T_r(Ex),Ex) = |a_2| = -a_2$ is a constant not depending on X. The functions $g_1(X)$ and $g_2(X)$ are given by

$$
\begin{aligned}
g_1(X) &:= l_2(X,T_l(X)) = \sqrt{\|X\|^2 - r^2} = \sqrt{x_1^2 + x_2^2 - 1}, \\
g_2(X) &:= r\phi = \phi.
\end{aligned}
$$

The angle ϕ can be obtained as

$$
\phi = \frac{\pi}{2} - \left(\arccos \frac{1}{\|X\|} + \arctan \frac{x_1}{x_2} \right)
$$

(see Figure 6.4), since

$$
\cos \alpha = \frac{1}{\|X\|} \qquad \text{and} \qquad \tan \delta = \frac{x_1}{x_2}.
$$

Therefore,

$$
g_2(X) = \frac{\pi}{2} - \arccos \left(\frac{1}{\sqrt{x_1^2 + x_2^2}} \right) - \arctan \left(\frac{x_1}{x_2} \right).
$$

Let $O \subset C_r(Ex)$ be an open convex set. Then $l_{2,B_C}(X,Ex) = g(X)$ is a convex function of X on the set O if and only if

$$
(Y - X)^T \cdot (\nabla g(Y) - \nabla g(X)) \geq 0 \qquad \forall X, Y \in O
$$

(see, for example, Rockafellar and Wets, 1998). The gradient of g satisfies $\nabla g(X) = \nabla g_1(X) + \nabla g_2(X)$, where the partial derivatives of g_1 and g_2 can be calculated as

$$\frac{\partial g_1}{\partial x_1} = \frac{x_1}{\sqrt{x_1^2 + x_2^2 - 1}},$$

$$\frac{\partial g_1}{\partial x_2} = \frac{x_2}{\sqrt{x_1^2 + x_2^2 - 1}},$$

$$\frac{\partial g_2}{\partial x_1} = \frac{1}{\sqrt{1 - \frac{1}{x_1^2 + x_2^2}}} \cdot \left(-\frac{1}{2}\left(x_1^2 + x_2^2\right)^{-\frac{3}{2}}\right) \cdot 2x_1 \;-\; \frac{1}{1 + \frac{x_1^2}{x_2^2}} \cdot \frac{1}{x_2}$$

$$= \frac{-x_1}{\sqrt{\frac{x_1^2 + x_2^2 - 1}{x_1^2 + x_2^2}} \cdot \left(x_1^2 + x_2^2\right)^{\frac{3}{2}}} \;-\; \frac{x_2}{x_1^2 + x_2^2}$$

$$= \frac{-x_1}{\sqrt{x_1^2 + x_2^2 - 1}\,\left(x_1^2 + x_2^2\right)} \;-\; \frac{x_2}{x_1^2 + x_2^2},$$

$$\frac{\partial g_2}{\partial x_2} = \frac{1}{\sqrt{1 - \frac{1}{x_1^2 + x_2^2}}} \cdot \left(-\frac{1}{2}\left(x_1^2 + x_2^2\right)^{-\frac{3}{2}}\right) \cdot 2x_2 \;-\; \frac{1}{1 + \frac{x_1^2}{x_2^2}} \cdot \left(-\frac{x_1}{x_2^2}\right)$$

$$= \frac{-x_2}{\sqrt{x_1^2 + x_2^2 - 1}\,\left(x_1^2 + x_2^2\right)} \;+\; \frac{x_1}{x_1^2 + x_2^2}.$$

Observe that $x_1^2 + x_2^2 \geq 1$ for all $X \in O$, since otherwise X would be located in the interior of the barrier $B_C = \{X \in \mathbb{R}^2 : \|X\| \leq 1\}$.

For two points $X = (x_1, x_2)^T$ and $Y = (y_1, y_2)^T$ in O, the scalar product $(Y - X)^T \cdot (\nabla g(Y) - \nabla g(X))$ can now be evaluated as

$$(Y - X)^T \cdot (\nabla g(Y) - \nabla g(X))$$

$$= (y_1 - x_1, y_2 - x_2)\begin{pmatrix} \dfrac{y_1\left(y_1^2 + y_2^2\right) - y_1}{\sqrt{y_1^2 + y_2^2 - 1}\,\left(y_1^2 + y_2^2\right)} - \dfrac{y_2}{y_1^2 + y_2^2} - \dfrac{x_1\left(x_1^2 + x_2^2\right) - x_1}{\sqrt{x_1^2 + x_2^2 - 1}\,\left(x_1^2 + x_2^2\right)} + \dfrac{x_2}{x_1^2 + x_2^2} \\[3ex] \dfrac{y_2\left(y_1^2 + y_2^2\right) - y_2}{\sqrt{y_1^2 + y_2^2 - 1}\,\left(y_1^2 + y_2^2\right)} + \dfrac{y_1}{y_1^2 + y_2^2} - \dfrac{x_2\left(x_1^2 + x_2^2\right) - x_2}{\sqrt{x_1^2 + x_2^2 - 1}\,\left(x_1^2 + x_2^2\right)} - \dfrac{x_1}{x_1^2 + x_2^2} \end{pmatrix}$$

$$= \frac{1}{y_1^2 + y_2^2}\left(\sqrt{y_1^2 + y_2^2 - 1}\,\left(y_1^2 + y_2^2 - x_1 y_1 - x_2 y_2\right) + x_1 y_2 - x_2 y_1\right)$$

$$+ \frac{1}{x_1^2 + x_2^2}\left(\sqrt{x_1^2 + x_2^2 - 1}\,\left(x_1^2 + x_2^2 - x_1 y_1 - x_2 y_2\right) - x_1 y_2 + x_2 y_1\right).$$

This expression is nonnegative for all X and Y in O satisfying $\|X\| = \|Y\|$ since in this case

$$(Y - X)^T \cdot (\nabla g(Y) - \nabla g(X))$$

$$= \frac{\sqrt{x_1^2 + x_2^2 - 1}}{x_1^2 + x_2^2} (y_1^2 + y_2^2 - x_1 y_1 - x_2 y_2 + x_1^2 + x_2^2 - x_1 y_1 - x_2 y_2)$$

$$= \frac{\sqrt{x_1^2 + x_2^2 - 1}}{x_1^2 + x_2^2} ((x_1 - y_1)^2 + (x_2 - y_2)^2) \geq 0.$$

If $\|X\| \neq \|Y\|$, a more sophisticated analysis of the above system is necessary. In the following, let

$$x := x_1^2 + x_2^2 \qquad \text{and} \qquad y := y_1^2 + y_2^2.$$

Recall that $x \geq 1$ and $y \geq 1$, since X and Y are not located in the interior of B_C. Moreover,

$$x_1 y_1 + x_2 y_2 = (x_1, x_2) \cdot (y_1, y_2)^T = \sqrt{x} \sqrt{y} \cos \rho$$

and

$$x_1 y_2 - x_2 y_1 = \sqrt{x} \sqrt{y} \sin \rho,$$

where ρ is the angle enclosed by the vectors X and Y (see Bronstein and Semendjajew, 1985).

Substituting this in the above formula, we obtain

$$(Y - X)^T \cdot (\nabla g(Y) - \nabla g(X))$$

$$= \frac{1}{y} \left(\sqrt{y-1} \; (y - \sqrt{x}\sqrt{y} \cos \rho) + \sqrt{x}\sqrt{y} \sin \rho \right)$$

$$+ \frac{1}{x} \left(\sqrt{x-1} \; (x - \sqrt{x}\sqrt{y} \cos \rho) - \sqrt{x}\sqrt{y} \sin \rho \right).$$

This implies that $(Y - X)^T \cdot (\nabla g(Y) - \nabla g(X)) \geq 0$ if and only if

$$\sqrt{x-1} + \sqrt{y-1} \geq \sqrt{x}\sqrt{y} \left(\left(\frac{\sqrt{y-1}}{y} + \frac{\sqrt{x-1}}{x} \right) \cos \rho + \left(\frac{1}{x} - \frac{1}{y} \right) \sin \rho \right),$$

which is equivalent to

$$\sqrt{x}\sqrt{y} \left(\sqrt{x-1} + \sqrt{y-1} \right) \geq \cos \rho \left(x\sqrt{y-1} + y\sqrt{x-1} \right) + \sin \rho \, (y - x).$$

$$\tag{6.5}$$

Obviously, the left-hand side of (6.5) is nonnegative, since $x, y \geq 1$. If the right-hand side of (6.5) is negative, then (6.5) is trivially satisfied. Otherwise, both sides of the inequality can be squared, implying the equivalent expression

$$xy \left(\sqrt{x-1} + \sqrt{y-1} \right)^2 \geq \left(\cos \rho \left(x\sqrt{y-1} + y\sqrt{x-1} \right) + \sin \rho \, (y - x) \right)^2.$$

$$\tag{6.6}$$

At this point Schwarz's inequality given by $\left(\sum_{i=1}^{n} a_i b_i\right)^2 \leq \left(\sum_{i=1}^{n} a_i^2\right) \cdot$ $\left(\sum_{i=1}^{n} b_i^2\right)$ for $a_i, b_i \in \mathbb{R}$, $i = 1, \ldots, n$ (see Bronstein and Semendjajew, 1985), can be applied to the right-hand side of (6.6), yielding

$$
(\cos^2 \rho + \sin^2 \rho) \cdot \left(\left(x\sqrt{y-1} + y\sqrt{x-1}\right)^2 + (y-x)^2\right)
$$
$$
\geq \left(\cos \rho \left(x\sqrt{y-1} + y\sqrt{x-1}\right) + \sin \rho(y - x)\right)^2.
$$

Using this inequality to bound the right-hand side of (6.6) and using the fact that $\cos^2 \rho + \sin^2 \rho = 1$ we can conclude that (6.6) is satisfied if

$$
xy\left(\sqrt{x-1} + \sqrt{y-1}\right)^2 \geq \left(x\sqrt{y-1} + y\sqrt{x-1}\right)^2 + (y-x)^2,
$$

or, equivalently, if

$$
xy\left(x + y - 2 + 2\sqrt{x-1}\sqrt{y-1}\right)
$$
$$
\geq x^2(y-1) + y^2(x-1) + 2xy\sqrt{x-1}\sqrt{y-1} + x^2 - 2xy + y^2.
$$

An easy calculation proves that both sides are equal, and thus this last inequality is satisfied for all $x, y \geq 1$.   □

In Lemma 6.2 we have shown that for a single existing facility $Ex \in \mathcal{E}x$ and one circular barrier the barrier distance $l_{2,B_C}(X, Ex)$ is a convex function of X on every open convex subset of $C_r(Ex)$ and of $C_l(Ex)$. Moreover, $l_{2,B_C}(X, Ex) = l_2(X, Ex)$ is a convex function of X on every open convex subset of $\mathcal{F} \backslash \text{shadow}_{l_2}(Ex)$. In the following, several existing facilities will be combined, and the objective function $f_B(X)$ will be analyzed over suitable subsets of $\mathcal{F}$.

Consider the grid $\mathcal{G}_{l_2}(B_C)$ defined by the boundaries of the shadows of all existing facilities, the construction lines $(\underline{0} - Ex)_{B_C}$, plus the boundary of the barrier region B_C:

$$
\mathcal{G}_{l_2}(B_C) := \bigcup_{Ex_m \in \mathcal{E}x} (\partial(\text{shadow}_{l_2}(Ex_m)) \cup (\underline{0} - Ex_m)_{B_C}) \cup \partial(B_C).
$$

The set $\mathcal{C}(\mathcal{G}_{l_2}(B_C))$ of cells induced by the grid $\mathcal{G}_{l_2}(B_C)$ is defined as the set of those (maximal) subsets of $\mathcal{F}$ that are bounded by curves or line segments of $\mathcal{G}_{l_2}(B_C)$ and whose interior is not intersected by any curve or line segment of $\mathcal{C}(\mathcal{G}_{l_2}(B_C))$; see Figure 6.5.

The overall number $|\mathcal{C}(\mathcal{G}_{l_2}(B_C))|$ of these cells can be estimated in the following way: Every existing facility $Ex \in \mathcal{E}x$ generates three half-lines that are part of the grid $\mathcal{G}_{l_2}(B_C)$, namely two half-lines starting at the left and right points of tangency, respectively, bounding the l_2-shadow of Ex, plus the construction line $(\underline{0} - Ex)_{B_C}$. The half-lines induced by the

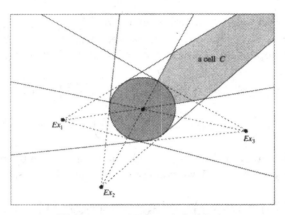

Figure 6.5. The grid $\mathcal{G}_{l_2}(B_C)$.

existing facilities in $\mathcal{E}x$ can intersect in at most $O\left(M^2\right)$ intersection points (where $M = |\mathcal{E}x|$ denotes the total number of existing facilities), and the boundary of B_C is intersected by these half-lines in not more than $O(M)$ different points. We construct a planar graph by approximating $\partial(B_C)$ by line segments between adjacent intersection points of $\partial(B_C)$ with the half-lines building the grid $\mathcal{G}_{l_2}(B_C)$. If we define a vertex at each intersection point of line segments in $\mathcal{G}_{l_2}(B_C)$, the number of vertices can be bounded by $O\left(M^2\right)$, and analogously the number of edges can be bounded by $O\left(M^2\right)$. Using Euler's formula for planar graphs (see Harary (1969) and the discussion of the grid $\mathcal{G}_{l_2}$ on page 61), the maximal number of cells in $\mathcal{C}(\mathcal{G}_{l_2}(B_C))$ can be bounded by $O\left(M^2\right) = O\left(|\mathcal{E}x|^2\right)$.

For the construction of the grid $\mathcal{G}_{l_2}(B_C)$ we first have to determine the points of tangency $T_r(Ex_m)$ and $T_l(Ex_m)$ for all existing facilities $Ex_m \in \mathcal{E}x$ according to equations (6.3). Then the half-lines defining the boundary of the l_2-shadow of an existing facility $Ex_m \in \mathcal{E}x$ can be found as $\{X \in \mathbb{R}^2 : X = T_r(Ex_m) + \lambda(T_r(Ex_m) - Ex_m), \lambda \geq 0\}$ and $\{X \in \mathbb{R}^2 : X = T_l(Ex_m) + \lambda(T_l(Ex_m) - Ex_m), \lambda \geq 0\}$, respectively.

All construction lines $(\underline{0} - Ex_m)_{B_C}$ are half-lines that can be easily calculated according to equation (6.4) for all $m = 1, \ldots, M$. In order to obtain a planar graph representing $\mathcal{G}_{l_2}(B_C)$ and having a vertex at every intersection point of line segments in $\mathcal{G}_{l_2}(B_C)$ (the boundary of B_C is again approximated by a piecewise linear curve), the intersection points of the $O(M)$ line segments of $\mathcal{G}_{l_2}(B_C)$ can be determined by an algorithm of Chazelle and Edelsbrunner (1992) in optimal $O(M \log M + k)$ time, where k denotes the total number of intersection points. If, in the worst case, $k = O\left(M^2\right)$, this implies a running time of the algorithm of $O\left(M^2\right)$. Alternatively, either a randomized algorithm as suggested by Clarkson and Shor (1989) with the same expected running time but requiring less space can be used to find the intersection points of all line segments in $\mathcal{G}_{l_2}(B_C)$, or the sweep-line algorithm of Bentley and Ottmann (1979) with a running time of $O((M + k) \log M)$ can be utilized.

In contrast to the definition of the visibility grid $\mathcal{G}_{l_2}$ on page 60, see Definition 5.2, the grid $\mathcal{G}_{l_2}(B_C)$ does not exclusively consist of line segments in $\mathbb{R}^2$. However, only the boundary of B_C contributes nonlinear pieces of arcs to $\mathcal{G}_{l_2}(B_C)$, which can easily be approximated by piecewise linear curves, and therefore the grid $\mathcal{G}_{l_2}(B_C)$ is computationally not more complex than the corresponding grid $\mathcal{G}_{l_2}$.

Moreover, in the case of a circular barrier the results developed in Section 5.1 for polyhedral barrier sets can be strengthened: It can be shown that the grid $\mathcal{G}_{l_2}(B_C)$ decomposes the feasible region into a finite number of cells such that the objective function of a problem of type $1/P/B_C/l_{2,B_C}/f$ convex is convex on every open convex subset of a cell in $\mathcal{C}(\mathcal{G}_{l_2}(B_C))$, even though the cells in $\mathcal{C}(\mathcal{G}_{l_2}(B_C))$ are not regions of constant intermediate points as defined in Section 5.1; see page 58.

Theorem 6.1. *Let $C \in \mathcal{C}(\mathcal{G}_{l_2}(B_C))$ be a cell of the grid $\mathcal{G}_{l_2}(B_C)$ for a problem of the type $1/P/B_C/l_{2,B_C}/f$ convex. Then the objective function $f_B(X)$ of problem (6.1) is convex on every open convex subset of C.*

Proof. Let $C \in \mathcal{C}(\mathcal{G}_{l_2}(B_C))$ be an arbitrary cell and let $O \subset C$ be an open convex subset of the cell C.

We first show that, as a consequence of Lemma 6.2, the distance $l_{2,B_C}(X, Ex_m)$ between a point $X \in O$ and every existing facility $Ex_m \in \mathcal{E}x$, $m = 1, \ldots, M$, is a convex function of X on O: Since C is a cell in $\mathcal{C}(\mathcal{G}_{l_2}(B_C))$, the interior of C is neither intersected by the boundary of the l_2-shadow of Ex nor by the construction line $(\underline{0} - Ex)_{B_C}$. Therefore, Ex either is l_2-visible from all points in $O \subset C$, which immediately implies the convexity of $l_{2,B_C}(X, Ex)$ on O, or C is a subset of one of the sets $C_r(Ex)$ and $C_l(Ex)$ of the l_2-shadow of Ex (see Figure 6.4), and the convexity of $l_{2,B_C}(X, Ex)$ on O follows from Lemma 6.2.

Hence

$$f_B = f(l_{2,B_C}(X, Ex_1), \ldots, l_{2,B_C}(X, Ex_M)) = f(\varphi_1, \ldots, \varphi_M),$$

where $\varphi_m : \mathbb{R}^2 \to \mathbb{R}$ with $\varphi_m(X) := l_{2,B_C}(X, Ex_m)$ are convex functions of X on O for every $m = 1, \ldots, M$. Therefore, the composition f_B of the convex and nondecreasing function f and the convex functions $\varphi_1, \ldots, \varphi_M$ is also a convex function of X on O (see Rockafellar and Wets, 1998). $\square$

As an example we will use the *Weber problem with a circular barrier* $1/P/B_C/\,l_{2,B_C}/\sum$. In this case, a positive weight $w_m = w(Ex_m)$ is identified with each existing facility $Ex_m \in \mathcal{E}x$ representing the demand of the facility Ex_m, and the problem $1/P/B_C/l_{2.B_C}/\sum$ is given by

$$\min \quad f_B(X) = \sum_{m=1}^{M} w_m l_{2.B_C}(X, Ex_m) \qquad (6.7)$$
$$\text{s.t.} \quad X \in \mathcal{F}.$$

The level sets

$$L_\le(z, f_B) := \{X \in \mathcal{F} \ : \ f_B(X) \le z\}$$

of an example problem with the Weber objective function (6.7) and with four existing facilities having equal weights $w_m = 1$, $m = 1, \ldots, 4$, are depicted in Figure 6.6. Since f_B is a convex function on every open convex subset O of a cell $C \in \mathcal{C}(\mathcal{G}_{l_2}(B_C))$. the intersection of a level set $L_\le(z, f_B)$ with an open convex set $O \subseteq C$ is a convex set for all $z \in \mathbb{R}$.

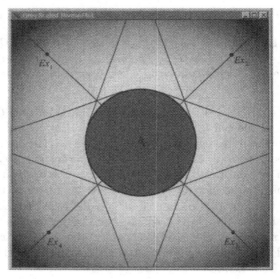

Figure 6.6. Level sets of a Weber problem (6.7) with four existing facilities having equal weights.

6.2 Algorithms and Heuristics

The convexity of the objective function on all open convex subsets of cells of the grid $\mathcal{G}_{l_2}(B_C)$ allows the development of efficient algorithms for the solution of problems of the type $1/P/B_C/l_{2.B_C}/f$ convex. The general idea for solving $1/P/B_C/l_{2.B_C}/f$ convex can be summarized as follows: First the grid $\mathcal{G}_{l_2}(B_C)$ is constructed and all of the at most $O(M^2) = O(|\mathcal{E}x|^2)$ cells are identified. Then the objective function f_B is minimized over each cell $C \in \mathcal{C}(\mathcal{G}_{l_2}(B_C))$ by some available method for convex optimization problems; see, for example, the textbooks of Bazaraa et al. (1993) and Polak (1997). A global minimum of $1/P/B_C/l_{2,B_C}/f$ convex can be calculated as the minimum of all these subproblems solved on the individual cells.

Algorithm 6.1. (Solution of $1/P/B_C/l_{2,B_C}/f$ convex)

Input: *Location problem $1/P/B_C/l_{2,B_C}/f$ convex.*

Step 1:   *Construct the grid $\mathcal{G}_{l_2}(B_C)$.*

Step 2:   *For all cells $C_k \in \mathcal{C}(\mathcal{G}_{l_2}(B_C))$, $k = 1, \ldots, K$ do:*

Find an optimal solution (or an approximation of the optimal solution) X_k^* of the convex optimization problem

$$\min\{f_\mathcal{B}(X) \, : \, X \in C_k\} \qquad (6.8)$$

by an available method.

Step 3:   *Determine an optimal solution*

$$X_{B_C}^* := \underset{k=1,\ldots,K}{\operatorname{argmin}} \, f_\mathcal{B}(X_k^*).$$

Output: *An optimal solution $X_{B_C}^*$ (or an approximation of an optimal solution $X_{B_C}^*$) of $1/P/B_C/l_{2,B_C}/f$ convex.*

Since the number K of cells of the grid $\mathcal{G}_{l_2}(B_C)$ can be bounded by $O\left(M^2\right)$ where $M = |\mathcal{E}x|$ denotes the total number of existing facilities, the time complexity of Algorithm 6.1 can be estimated as $O\left(M^2 \cdot T\right)$, where $O(T)$ is the time complexity for solving the convex subproblems (6.8). Note that the time complexity of constructing the grid $\mathcal{G}_{l_2}(B_C)$ in Step 1 of the algorithm is dominated by the time complexity of Step 2; cf. the discussion of the construction of $\mathcal{G}_{l_2}(B_C)$ on page 94.

Algorithm 6.1 was implemented based on the Library of Location Algorithms LoLA (see Hamacher et al. (1999a) and the diploma thesis of Ochs (1998)). In this implementation all the convex subproblems are solved by adapting the method of Hooke and Jeeves (see Hooke and Jeeves, 1961).

To compare a special case of Algorithm 6.1 with the results of Katz and Cooper (1981) for Weber problems with a circular barrier, a reference problem introduced in Katz and Cooper (1981) was used. In this problem, five existing facilities each with weight 1 are given at the coordinates $Ex_1 = (-8.0, -6.0)^T$, $Ex_2 = (-7.0, 13.0)^T$, $Ex_3 = (-1.0, -5.0)^T$, $Ex_4 = (6.6, -0.5)^T$, $Ex_5 = (4.4, 10.0)^T$, and one circular barrier with radius 2 centered at the origin $\underline{0} = (0, 0)^T$ is located within the considered region. The optimal solution was approximated by Algorithm 6.1 at the point $X_1 = (-1.18602, 2.06044)^T$ with an objective value of $z_1 = 48.2548$. This result slightly improves the solution $X_2 = (-1.2016, 2.0776)^T$ with $z_2 = 48.2560$ as found in Butt and Cavalier (1996), who approximated the circular barrier by a polyhedral set and used a heuristic algorithm that iteratively solves unconstrained location problems. The solution determined in the original work of Katz and Cooper (1981) at the point $X_3 = (-0.08130, 2.4833)^T$ with an objective value of $z_3 = 48.3524$ turned out to be a local minimum located relatively far from the approximated optimum at the point $X_1 = (-1.18602, 2.06044)^T$.

In addition to the exact solution procedure given in Algorithm 6.1, the following heuristic approach was implemented (see Ochs (1998)): As in Algorithm 6.1, first the grid $\mathcal{G}_{l_2}(B_C)$ is constructed. The planar graph representing $\mathcal{G}_{l_2}(B_C)$ with vertices at all intersection points of line segments in $\mathcal{G}_{l_2}(B_C)$ is determined as discussed above (see page 94). In contrast to Algorithm 6.1, not all the cells are investigated during the following heuristic procedure. Instead, the objective value is first determined only at the at most $O(M^2)$ vertices of the graph representing $\mathcal{G}_{l_2}(B_C)$. These vertices are then stored in a list that is sorted according to the objective values of the list's entries. For a given percentage $p \in (0.100)$, the best $p\%$ of vertices of this list are selected for further investigation. The selected points are used as starting points for the method of Hooke and Jeeves applied to the minimization problem

$$\min \quad f_B(X)$$
$$\text{s.t.} \quad X \in \mathcal{F}.$$

where the constraint to cells of $\mathcal{G}_{l_2}(B_C)$ as imposed in Algorithm 6.1 is relaxed to the weaker feasibility constraint $X \in \mathcal{F}$. Therefore, several cells may be visited during one application of the method of Hooke and Jeeves. Moreover, the objective function f_B is in general nonconvex on the whole feasible region $\mathcal{F}$. Since the objective function is convex on every cell of $\mathcal{G}_{l_2}(B_C)$, we may expect that the method of Hooke and Jeeves converges to a minimum in a cell that contains a global minimum of $1/P/B_C/l_{2,B_C}/f$ convex. Therefore, we stop the minimization of a subproblem whenever during the procedure a cell of $\mathcal{G}_{l_2}(B_C)$ is reached that was already visited. In this case we proceed with the next selected point and use it as the starting point for a new application of the method of Hooke and Jeeves.

Algorithm 6.2. (Heuristic Solution of $1/P/B_C/l_{2,B_C}/f$ convex)

Input: *Location problem* $1/P/B_C/l_{2,B_C}/f$ *convex; parameter* $p \in (0, 100)$ *specifying the percentage of starting points considered.*

Step 1:   *Construct the grid* $\mathcal{G}_{l_2}(B_C)$ *and the corresponding planar graph representing* $\mathcal{G}_{l_2}(B_C)$.

Step 2:   *Determine the objective value for all intersection points in* $\mathcal{G}_{l_2}(B_C)$ *and store these points in a list sorted according to their objective values. Select the best p percent of points in this list for further investigation.*

Step 3:   *For all points Y in the list of starting points determined in Step*
           2, do:

           Apply the method of Hooke and Jeeves to the problem

$$\min\{f_B(X) \: : \: X \in \mathcal{F}\} \tag{6.9}$$

           with starting point Y. Add all cells visited during the pro-
           cedure to a list of cells either until a cell is reached that
           was already visited by an earlier run of the method of
           Hooke and Jeeves, or until the method converges to a lo-
           cal minimum of (6.9).

Step 4:   *Determine a minimizer of the solutions of the subproblems*
           solved in Step 3.

Output: *A heuristic solution X_{B_C} of $1/P/B_C/l_{2,B_C}/f$ convex.*

The heuristic Algorithm 6.2 was tested against the exact Algorithm 6.1 for
the case of Weber problems $1/P/B_C/l_{2,B_C}/\sum$ (see Ochs, 1998). In all of
the 54 randomly generated examples with a set of 20 to 60 existing facilities
an optimal solution was approximated by Algorithm 6.2. The percentage p
of investigated starting points was set to $p = 10\%$ in these tests.

7
Weber Problems with a Line Barrier

The special case of almost linear barriers in the plane that have only a finite number of passages is frequently encountered in practice. Line barriers with passages may be used to model rivers, border lines, highways, mountain ranges, or, on a smaller scale, conveyer belts in industrial plants. In all these examples trespassing is allowed only through a finite number of passages. Disregarding these types of barriers may lead to bad locational decisions, since long and almost linear barriers have a particularly big impact on travel distances and travel times.

If a line barrier is modeled as a very long and narrow polyhedral set, the results developed in Chapter 5 can also be applied in this case. We will, however, use a different approach in this chapter based on the approximation of these types of barriers by lines having a finite number of passages through which traveling is allowed. Using this model, the Weber objective function will be further analyzed for a large class of metrics (including, for example, all l_p metrics, $p \in [1, \infty]$). Moreover, we will show that the number of convex (unconstrained) subproblems that would have to be solved in Algorithm 5.2 (see Section 5.3) can be significantly reduced with this approach. As in Section 5.5 (see page 77) the following results can be generalized to objective functions f that are homomorphisms satisfying (5.16). Parts of the results presented in this chapter can also be found in Klamroth (2001a).

Planar location problems with a line barrier can be modeled as follows: Let

$$L := \left\{ (x, y)^T \in \mathbb{R}^2 \ : \ ax + by = c \right\}$$

with $a, b, c \in \mathbb{R}$ be a line in $\mathbb{R}^2$ and let

$$\{P_n \in L \; : \; n \in \mathcal{N} = \{1, \ldots, N\}\}$$

be a finite set of points on L.

Definition 7.1. *For a line L and a set of points P_i, $i = 1, \ldots, N$, on L,*

$$\mathcal{B}_L := L \setminus \{P_1, \ldots, P_N\}$$

is called a line barrier with passages *or, shortly, a* line barrier.

The feasible region $\mathcal{F}$ for new locations is defined as the union of the two closed half-planes $\mathcal{F}^1$ and $\mathcal{F}^2$ on both sides of $\mathcal{B}_L$; i.e., $\mathcal{F}^1 \cup \mathcal{F}^2 = \mathbb{R}^2$, since the line L belongs to both half-planes $\mathcal{F}^1$ and $\mathcal{F}^2$. Since all results can easily be transferred to the case that the line barrier has a finite width, for simplification this model will be used in the following, although a new location placed directly on the barrier is generally not allowed in practice. Consequently, if a point lies directly on the line, we will nevertheless assign the point X to only one of the half-planes $\mathcal{F}^1$ or $\mathcal{F}^2$, but never to both half-planes at the same time.

If a metric d is given for the unconstrained problem, the barrier distance function $d_{\mathcal{B}_L}$ for a problem of the type $1/P/(\mathcal{B} = 1\,line)/d_{\mathcal{B}}/\sum$, or $1/P/\mathcal{B}_L/d_{\mathcal{B}_L}/\sum$ for brevity, is defined by

$$d_{\mathcal{B}_L}(X, Y) := \min_{\substack{k \in \mathbb{N} \\ T_1, \ldots, T_k \in \mathcal{F}}} \sum_{i=1}^{k-1} d(T_i, T_{i+1})$$

with $T_1 = X$, $T_k = Y$, and $T_i \notin \mathcal{B}_L$ $(i = 1, \ldots, k)$ such that the straight line segment connecting T_i and T_{i+1} does not cross $\mathcal{B}_L$. Note that this definition differs from Definition 2.2, since a line barrier as defined in Definition 7.1 has empty interior. However, trespassing through $\mathcal{B}_L$ is allowed only at the passage points $\{P_1, \ldots, P_N\}$.

We obtain the following simplified description of $d_{\mathcal{B}_L}$:

Lemma 7.1. *Let d be a metric induced by a norm and let $i \in \{1, 2\}$. Then*

$$d_{\mathcal{B}_L}(X, Y) = \begin{cases} d(X, Y) & \text{if } X, Y \in \mathcal{F}^i, \\ d(X, P_{n_{X,Y}}) + d(P_{n_{X,Y}}, Y) & \text{for some } n_{X,Y} \in \mathcal{N} \text{ otherwise.} \end{cases}$$

Proof. Since $\mathcal{F}^i$ is a convex set $(i \in \{1, 2\})$ the straight line segment connecting two points X and Y in the same half-plane $\mathcal{F}^i$ is a d-shortest permitted X-Y path (see Lemma 2.2 on page 19). Furthermore, due to the triangle inequality, every d-shortest permitted path from a point $X \in \mathcal{F}^1$ to another point $Y \in \mathcal{F}^2$ has to pass through exactly one of the passages P_n, $n \in \mathcal{N}$. Observe that this passage point can be interpreted as an intermediate point according to Definition 2.5 on page 33. Thus $d_{\mathcal{B}_L}(X,Y) = d(X, P_{n_{X,Y}}) + d(P_{n_{X,Y}}, Y)$ holds for some $n_{X,Y} \in \mathcal{N}$, where the passage $P_{n_{X,Y}}$ depends on the coordinates of X and Y. $\qquad\qquad\square$

Furthermore, a finite number of existing facilities $Ex_m^i \in \mathcal{F}^i$, $m \in \mathcal{M}^i :=$ $\{1, \ldots, M^i\}$, is given in each half-plane $\mathcal{F}^i$, $i = 1, 2$; see Figure 7.1. In the case of a Weber objective function, we assume that a nonnegative weight $w_m^i := w\left(Ex_m^i\right) \in \mathbb{R}_+$ is associated with each existing facility Ex_m^i, representing the demand of Ex_m^i.

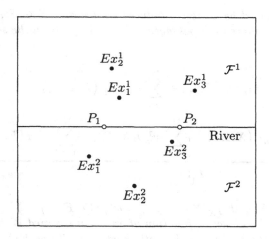

Figure 7.1. An example problem modeling a river with two bridges.

Then the *Weber problem with a line barrier* $1/P/\mathcal{B}_L/d_{\mathcal{B}_L}/\sum$ can be formulated as

$$\min \quad f_{\mathcal{B}}(X) = \sum_{m=1}^{M^1} w_m^1 d_{\mathcal{B}_L}\left(X, Ex_m^1\right) + \sum_{m=1}^{M^2} w_m^2 d_{\mathcal{B}_L}\left(X, Ex_m^2\right)$$

$$\text{s.t.} \quad X \in \mathcal{F}.$$

The general idea for solving problems of the type $1/P/\mathcal{B}_L/d_{\mathcal{B}_L}/\sum$ can be summarized as follows: Assume that an optimal solution of the problem is located in the half-plane $\mathcal{F}^1$. Then the passages channel the flow from the existing facilities in $\mathcal{F}^2$ to the new location in $\mathcal{F}^1$. Interpreting these passages as artificial facilities carrying the weights of the assigned existing facilities in $\mathcal{F}^2$, the new location can be retrieved as the solution of an unconstrained Weber problem in the half-plane $\mathcal{F}^1$. Finding the relevant subset of passage points and their respective weights is a combinatorial problem that will be discussed in Sections 7.2 and 7.3 below.

In the following we will derive the theoretical basis for this approach for a large class of Weber problems with distance functions, including, for example, the class of all l_p metrics, $p \in [1, \infty]$.

We can use Lemma 7.1 as we used Lemma 5.3 and Corollary 5.3 to rewrite the objective function for a point $X \in \mathcal{F}^i$. Yet in this case we are in a position to identify the passage points as the only candidates for intermediate points and replace the nonconvex barrier distance function $d_{\mathcal{B}_L}$ entirely by the given metric d. Moreover, a decomposition of the feasible region based on the visibility grid $\mathcal{G}_d$ can be avoided in this case, since the passage points $P_1, \ldots, P_N$ are d-visible from all points in $\mathcal{F}^1$ and in $\mathcal{F}^2$.

Corollary 7.1. *Let d be a metric induced by a norm, $X \in \mathcal{F}^i$, and $i, j \in \{1, 2\}$, $i \neq j$. Then there exist passages $P_{n_1}, \ldots, P_{n_{M^j}}$ from the set $\{P_1, \ldots, P_N\}$ such that*

$$
\begin{aligned}
f_{\mathcal{B}}(X) &= \sum_{m=1}^{M^i} w_m^i d_{\mathcal{B}_L}\left(X, Ex_m^i\right) + \sum_{m=1}^{M^j} w_m^j d_{\mathcal{B}_L}\left(X, Ex_m^j\right) \\
&= \underbrace{\sum_{m=1}^{M^i} w_m^i d\left(X, Ex_m^i\right) + \sum_{m=1}^{M^j} w_m^j d(X, P_{n_m})}_{=:f_X^i(X)} + \underbrace{\sum_{m=1}^{M^j} w_m^j d\left(Ex_m^j, P_{n_m}\right)}_{=:g_X^j}.
\end{aligned} \tag{7.1}
$$

Here, $f_X^i : \mathcal{F}^i \to \mathbb{R}$ with $f_X^i(Y) := \sum_{m=1}^{M^i} w_m^i d\left(Y, Ex_m^i\right) + \sum_{m=1}^{M^j} w_m^j d(Y, P_{n_m})$, $Y \in \mathcal{F}^i$, is the objective function of an unconstrained Weber problem $1/P/\bullet/d/\sum$ in the half-plane $\mathcal{F}^i$ with existing facilities $Ex_1^i, \ldots, Ex_{M^i}^i$, $P_1, \ldots, P_N$; and $g_X^j := \sum_{m=1}^{M^j} w_m^j d\left(Ex_m^j, P_{n_m}\right)$ can be interpreted as a constant, since it depends only implicitly on X via the selection of the passage points P_{n_m}, $m = 1, \ldots, M^j$.

We will say that $P_{n_m} \in \{P_1, \ldots, P_N\}$ is *assigned* to $Ex_m \in \mathcal{E}x$ if a d-shortest path from Ex_m to a point in the opposite half-plane passes through P_{n_m}.

Note that, in analogy to Lemma 5.5 on page 65, we obtain that

$$
f_X^i(Y) + g_X^j \geq f_Y^i(Y) + g_Y^j = f_{\mathcal{B}_L}(Y) \qquad \forall X, Y \in \mathcal{F}^i, \tag{7.2}
$$

since the assignment of passages to the existing facilities in the opposite half-plane $\mathcal{F}^j$ used in the reformulation of the objective function for a point $X \in \mathcal{F}^i$ in Corollary 7.1 is feasible for all points $Y \in \mathcal{F}^i$, but it is in general not the optimal assignment ($i, j \in \{1, 2\}$, $i \neq j$).

Corollary 7.1 can be used to calculate the set of optimal solutions $\mathcal{X}_{\mathcal{B}}^*$ of the constrained Weber problem $1/P/\mathcal{B}_L/d_{\mathcal{B}_L}/\sum$ from the sets of optimal solutions of a finite number of unconstrained Weber problems of the type $1/P/\bullet/d/\sum$. For this purpose, let d be a metric induced by a norm such that the *convex hull property*

$$\mathcal{X}^* \subseteq \text{conv}\{Ex_m \ : \ m \in \mathcal{M}\} \tag{7.3}$$

holds for every unconstrained location problem of the type $1/P/\bullet/d/\sum$.

Theorem 7.1. *Let d be a metric induced by a norm and let $1/P/\bullet/d/\sum$ be a Weber problem such that the convex hull property (7.3) is satisfied.*
Then every optimal solution $X_{\mathcal{B}}^* \in \mathcal{X}_{\mathcal{B}}^*$ of $1/P/\mathcal{B}_L/d_{\mathcal{B}_L}/\sum$ is also an optimal solution of the corresponding unconstrained Weber problem $1/P/\bullet/d/\sum$ with existing facilities $Ex_1^i, \dots, Ex_{M^i}^i, P_1, \dots, P_N$ in $\mathcal{F}^i$ and objective function $f_{X_{\mathcal{B}}^*}^i$ ($i \in \{1,2\}$).

Proof. Let $X_{\mathcal{B}}^* \in \mathcal{F}^i$ be an optimal solution of $1/P/\mathcal{B}_L/d_{\mathcal{B}_L}/\sum$. From Corollary 7.1 we know that

$$f_{\mathcal{B}}(X_{\mathcal{B}}^*) = f_{X_{\mathcal{B}}^*}^i(X_{\mathcal{B}}^*) + g_{X_{\mathcal{B}}^*}^j.$$

Here $g_{X_{\mathcal{B}}^*}^j$ is constant and $f_{X_{\mathcal{B}}^*}^i(Y)$ is the objective function of the corresponding unconstrained Weber problem with existing facilities $Ex_1^i, \dots, Ex_{M^i}^i, P_1, \dots, P_N$ in $\mathcal{F}^i$, and with a set of optimal solutions $\mathcal{X}^* \subseteq \text{conv}\{Ex_m^i, P_n \ : \ m \in \mathcal{M}^i, n \in \mathcal{N}\}$.
    Assume that $\exists Y \in \mathcal{X}^*$ with $f_{X_{\mathcal{B}}^*}^i(Y) < f_{X_{\mathcal{B}}^*}^i(X_{\mathcal{B}}^*)$. Since $Y \in \mathcal{F}^i$, inequality (7.2) implies that $f_{\mathcal{B}}(Y) \leq f_{X_{\mathcal{B}}^*}^i(Y) + g_{X_{\mathcal{B}}^*}^j < f_{X_{\mathcal{B}}^*}^i(X_{\mathcal{B}}^*) + g_{X_{\mathcal{B}}^*}^j = f_{\mathcal{B}}(X_{\mathcal{B}}^*)$, which is a contradiction to the optimality of $X_{\mathcal{B}}^*$. □

Theorem 7.1 is stronger than the results of Theorems 5.2 and 5.5, since the construction of the grid $\mathcal{G}_d$ is not required in this case, and since only the passage points $P_1, \dots, P_N$ are considered as candidates for intermediate points for those existing facilities that are not d-visible from a given solution.

The assumption of Theorem 7.1 based on the convex hull property (7.3) is satisfied for a large class of metrics. A well-known example is the class of l_p metrics with $1 < p < \infty$; see Juel and Love (1983). On the other hand, there exist some metrics such as the l_1 and l_∞ metrics for which only the *weak convex hull property*

$$\exists X^* \in \mathcal{X}^* \quad \text{with} \quad X^* \in \text{conv}\{Ex_m \ : \ m \in \mathcal{M}\} \tag{7.4}$$

holds for $1/P/\bullet/d/\sum$. A similar result to that given in Theorem 7.1 can be proven in this case:

Corollary 7.2. *Let d be a metric induced by a norm and let $1/P/\bullet/d/\sum$ be a Weber problem such that the weak convex hull property (7.4) is satisfied.*
Then at least one optimal solution $X_{\mathcal{B}}^* \in \mathcal{X}_{\mathcal{B}}^*$ of $1/P/\mathcal{B}_L/d_{\mathcal{B}_L}/\sum$ exists that is also an optimal solution of an analogous unconstrained Weber problem $1/P/\bullet/d/\sum$ in $\mathcal{F}^i$ with existing facilities $Ex_1^i, \dots, Ex_{M^i}^i, P_1, \dots, P_N$ ($i \in \{1,2\}$) and objective function $f_{X_{\mathcal{B}}^*}^i$.

In the following we restrict our discussion to such metrics d induced by norms such that the convex hull property (7.3) holds for every unconstrained location problem of the type $1/P/\bullet/d/\sum$. However, all results can be easily transferred to the case that only the weak convex hull property (7.4) holds for $1/P/\bullet/d/\sum$. In the first case the complete set of optimal locations $\mathcal{X}_\mathcal{B}^*$ of the constrained problem $1/P/\mathcal{B}_L/d_{\mathcal{B}_L}/\sum$ can be determined, whereas in the latter case at least one optimal solution $X_\mathcal{B}^* \in \mathcal{X}_\mathcal{B}^*$ is found. Juel and Love (1983) and Wendell and Hurter (1973) showed that at least the weak convex hull property (7.4) is satisfied for *all* unconstrained Weber problems with a metric induced by a norm.

Theorem 7.1 can be used to restrict the location of the set of optimal solutions $\mathcal{X}_\mathcal{B}^*$ of $1/P/\mathcal{B}_L/d_{\mathcal{B}_L}/\sum$ to the union of the convex hulls of the existing facilities together with the set of passages on either side of $\mathcal{B}_L$.

Theorem 7.2. *Let d be a metric induced by a norm and let $1/P/\bullet/d/\sum$ be a Weber problem for which the convex hull property (7.3) is satisfied. Then*

$$\mathcal{X}_\mathcal{B}^* \subseteq \operatorname{conv}\{Ex_m^1, P_n \,:\, m \in \mathcal{M}^1,\, n \in \mathcal{N}\}$$
$$\cup \operatorname{conv}\{Ex_m^2, P_n \,:\, m \in \mathcal{M}^2,\, n \in \mathcal{N}\}$$

holds for problems of type $1/P/\mathcal{B}_L/d_{\mathcal{B}_L}/\sum$.

Proof. Let $X_\mathcal{B}^* \in \mathcal{F}^i$, $i \in \{1,2\}$, be an optimal solution of $1/P/\mathcal{B}_L/d_{\mathcal{B}_L}/\sum$. Then Theorem 7.1 implies that $X_\mathcal{B}^*$ is an optimal solution of the corresponding unconstrained Weber problem $1/P/\bullet/d/\sum$ with existing facilities $Ex_1^i,\dots,Ex_{M^i}^i, P_1,\dots,P_N$, and objective function $f_{X_\mathcal{B}^*}^i$. Since this unconstrained Weber problem satisfies the convex hull property (7.3), we can conclude that $X_\mathcal{B}^* \in \operatorname{conv}\{Ex_m^i, P_n : m \in \mathcal{M}^i, n \in \mathcal{N}\}$. □

Note that in the case of a line barrier this result is stronger than the result of Theorem 5.3 on page 69 since the iterative convex hull may be unbounded in the case of an unbounded barrier.

Theorem 7.2 can be easily transferred to the case that only the weak convex hull property (7.4) is true for the unconstrained Weber problem $1/P/\bullet/d/\sum$:

Theorem 7.3. *Let d be a metric induced by a norm and let $1/P/\bullet/d/\sum$ be a Weber problem such that the weak convex hull property (7.4) is satisfied. Then for all problems of type $1/P/\mathcal{B}_L/d_{\mathcal{B}_L}/\sum$, there exists an optimal solution $X_\mathcal{B}^* \in \mathcal{X}_\mathcal{B}^*$ with*

$$X_\mathcal{B}^* \in \operatorname{conv}\{Ex_m^1, P_n \,:\, m \in \mathcal{M}^1,\, n \in \mathcal{N}\}$$
$$\cup \operatorname{conv}\{Ex_m^2, P_n \,:\, m \in \mathcal{M}^2,\, n \in \mathcal{N}\}.$$

Proof. Let $f_\mathcal{B}(X_\mathcal{B}^*)$ be the optimal objective function value of $1/P/\mathcal{B}_L/d_{\mathcal{B}_L}/\sum$ and without loss of generality let $X_\mathcal{B}^* \in \mathcal{F}^1$. Then $f_\mathcal{B}(X_\mathcal{B}^*) =$

$f^1_{X^*_B}(X^*_B) + g^2_{X^*_B}$, where $f^1_{X^*_B}$ is the objective function of an unconstrained problem $1/P/\bullet/d/\sum$ with existing facilities $Ex^1_1, \ldots, Ex^1_{M^1}, P_1, \ldots, P_N$. From the assumptions it follows that $\exists\ X^* \in \text{conv}\{Ex_m, P_n\ :\ m \in \mathcal{M},\ n \in \mathcal{N}\}$ with $f^1_{X^*_B}(X^*) = f^1_{X^*_B}(X^*_B)$, and thus $f_B(X^*) \leq f_B(X^*_B)$.   □

Note that in some special cases, for example in the case of a horizontal line barrier and l_1-distances, Theorem 7.2 can be applied even though the corresponding unconstrained problems don not satisfy the convex hull property (7.3). However, it can be shown (see Love and Morris, 1975) that the set of optimal solutions of problems of type $1/P/\bullet/l_1/\sum$ is always contained in the *rectangular hull* of the existing facilities, i.e., in the smallest axes-parallel rectangle circumscribing all the existing facilities. Since this rectangle is completely contained in one of the half-planes $\mathcal{F}^1$ or $\mathcal{F}^2$ if the set of existing facilities is a combination of facilities in one half-plane and of passage points, and if the line barrier is defined by a horizontal line, Theorem 7.2 also generalizes to this case.

Unfortunately, it is not possible to restrict $\mathcal{X}^*_B$, for example, to that half-plane with the higher total weight, as one may conjecture intuitively. This can easily be seen in the example given in Figure 7.2, where the weights of all sites are chosen equal to one and distances are measured by the Euclidean metric. The optimal solution X^*_B is located in $\mathcal{F}^1$, whereas the higher total weight is assumed in $\mathcal{F}^2$.

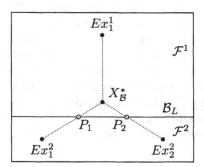

Figure 7.2. An example problem with $d = l_2$ and weights $w^1_1 = w^2_1 = w^2_2 = 1$ where the unique optimal solution X^*_B is located on the opposite side of the line barrier from the existing facilities with the higher total weight.

It is also easy to construct examples where $\mathcal{X}^*_B$ and $\mathcal{X}^*$, the set of optimal solutions of the corresponding problem disregarding the barrier $\mathcal{B}_L$, are located on opposite sides of $\mathcal{B}_L$; for an example see Figure 7.3.

Theorem 7.1 and Corollary 7.2 imply a very simple algorithm to solve problems of the type $1/P/\mathcal{B}_L/d_{\mathcal{B}_L}/\sum$. The basic idea is to check, for all existing facilities in one half-plane $\mathcal{F}^j$, all possible assignments of passages to the opposite half-plane $\mathcal{F}^i$ and to determine the sets of optimal solutions of the corresponding unconstrained location problems in $\mathcal{F}^i$. Given N

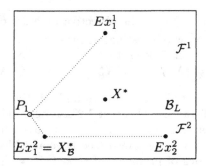

Figure 7.3. An example problem with $d = l_2$ and weights $w_1^1 = w_1^2 = w_2^2 = 1$ where the unique optimal solution $X_{\mathcal{B}}^*$ is located on the opposite side of the line barrier from the unique optimal solution X^* of the corresponding unconstrained problem for the same existing facilities.

passage points and M^j existing facilities in $\mathcal{F}^j$, this yields a total number of N^{M^j} subproblems that have to be solved in the half-plane $\mathcal{F}^i$, and this procedure must be carried out for both half-planes.

A necessary condition for the correctness of this procedure is that the convex hull property (7.3) or the weak convex hull property (7.4) be satisfied for the unconstrained location problem $1/P/\bullet/d/\sum$, entailing Theorem 7.1 or Corollary 7.2 to be applicable.

A more efficient algorithm that is also based on Theorem 7.1 and Corollary 7.2, i.e., the reduction of the nonconvex optimization problem $1/P/\mathcal{B}_L/d_{\mathcal{B}_L}/\sum$ to a finite set of convex optimization problems, will be developed in the following sections. It will be shown that only a small number of unconstrained location problems of the form $1/P/\bullet/d/\sum$ must be considered to obtain a candidate set for the set of optimal solutions of the original problem $1/P/\mathcal{B}_L/d_{\mathcal{B}_L}/\sum$. The time complexity of all the deduced algorithms discussed in the following is polynomial in the number of existing facilities if the number of passage points is finite and fixed.

7.1 Line Barriers with One Passage

First consider the case that only one passage P_1 allows trespassing through $\mathcal{B}_L$ and let $X \in \mathcal{F}^i$ and $i, j \in \{1, 2\}$, $i \neq j$. Then every path between an existing facility $Ex_m^j \in \mathcal{F}^j$ and X has to pass through P_1. Thus the objective function for the point X can be written as

$$f_{\mathcal{B}}(X) = \underbrace{\sum_{m=1}^{M^i} w_m^i d\left(X, Ex_m^i\right) + \left(\sum_{m=1}^{M^j} w_m^j\right) d(X, P_1)}_{=:f_X^i(X)} + \underbrace{\sum_{m=1}^{M^j} w_m^j d\left(P_1, Ex_m^j\right)}_{=:g_X^j}.$$

Observe that for all $X, Y \in \mathcal{F}^i$, $f_X^i = f_Y^i$ is the objective function of a Weber problem with existing facilities $Ex_1^i, \ldots, Ex_{M^i}^i, P_1$.

Since in this case the weight of P_1 in the corresponding unconstrained problem with respect to Theorem 7.1 is the sum of the weights of all existing facilities in one half-plane, $w(P_1) = \sum_{m=1}^{M^j} w_m^j$, the optimal new location can be restricted to the half-plane with the higher total weight (unless it is equal for both half-planes). Therefore, an unconstrained Weber problem has to be solved twice at most, once for each half-plane.

The time complexity of this procedure is $O(T)$, where $O(T)$ is the time complexity of the corresponding unconstrained Weber problem. If, for example, a problem with l_1 or l_∞ distances is given, the unconstrained Weber problem can be solved in time $O(T) = O(M \log M)$, where $M = M^1 + M^2$ denotes the total number of existing facilities; see Hamacher (1995).

7.2 Line Barriers with Two Passages

Under the more realistic assumption that more than one passage allows trespassing through $\mathcal{B}_L$, the number of unconstrained Weber problems that have to be solved to obtain the set of optimal solutions can be reduced considerably compared to the straightforward approach described above.

For both half-planes $\mathcal{F}^i$, $i = 1, 2$, the difference of distances $D^i(m)$ between an existing facility $Ex_m^i \in \mathcal{F}^i$ and the two passages P_1 and P_2 is defined as

$$D^i(m) := d\left(Ex_m^i, P_1\right) - d\left(Ex_m^i, P_2\right), \qquad m \in \mathcal{M}^i.$$

Without loss of generality assume that the existing facilities are ordered such that $D^i(1) \leq \cdots \leq D^i(M^i)$. Furthermore, let $j \in \{1, 2\}$ with $j \neq i$ be the index of the opposite half-plane $\mathcal{F}^j$. A d-shortest permitted path SP from an existing facility $Ex_m^j \in \mathcal{F}^j$ to a point $X \in \mathcal{F}^i$ passes through one of the passages P_1 and P_2 depending on the following condition (see Figure 7.4):

$$
\begin{aligned}
P_1 \in SP &\iff d\left(Ex_m^j, P_1\right) + d(P_1, X) < d\left(Ex_m^j, P_2\right) + d(P_2, X), \\
P_2 \in SP &\iff d\left(Ex_m^j, P_1\right) + d(P_1, X) > d\left(Ex_m^j, P_2\right) + d(P_2, X).
\end{aligned}
\tag{7.5}
$$

In the case that $d\left(Ex_m^j, P_1\right) + d(P_1, X) = d\left(Ex_m^j, P_2\right) + d(P_2, X)$, a d-shortest permitted path may pass through either passage P_1 or P_2.

Using the definition of the difference of distances D_m^i, (7.5) can be reformulated as

$$
\begin{aligned}
P_1 \in SP &\iff D^j(m) < d(P_2, X) - d(P_1, X), \\
P_2 \in SP &\iff D^j(m) > d(P_2, X) - d(P_1, X).
\end{aligned}
\tag{7.6}
$$

Note that the set of points in $\mathcal{F}^i$ for which both passages lie on a d-shortest permitted path to an existing facility $Ex_m^j \in \mathcal{F}^j$ is in general

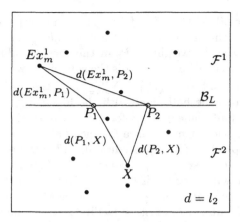

Figure 7.4. The d-shortest permitted path from Ex_m^1 to X depends on $D^1(m)$.

not linear. Depending on the distance function d, it defines, for example, a hyperbolic curve; cf. Figure 5.1 on page 59. However, the computation of the regions of constant intermediate points as introduced in Section 5.1 (see Definition 5.1 on page 58), where the passage points assigned to the existing facilities do not change, can be avoided by defining

$$k := \max \left\{ m \in \{0, 1, \ldots, M^j\} \ : \ D^j(m) < d(P_2, X) - d(P_1, X) \right\}. \quad (7.7)$$

Then, as in (7.1), the value of the objective function $f_{\mathcal{B}}(X)$ for the point $X \in \mathcal{F}^i$ can be evaluated as

$$f_{\mathcal{B}}(X) = f_k^i(X) + g_k^j, \quad (7.8)$$

where

$$f_k^i(Y) \ := \ \sum_{m=1}^{M^i} w_m^i d\left(Y, Ex_m^i\right) + \left(\sum_{m=1}^{k} w_m^j\right) d(Y, P_1)$$

$$+ \left(\sum_{m=k+1}^{M^j} w_m^j\right) d(Y, P_2), \quad (7.9)$$

$$g_k^j \ := \ \sum_{m=1}^{k} w_m^j d\left(P_1, Ex_m^j\right) + \sum_{m=k+1}^{M^j} w_m^j d\left(P_2, Ex_m^j\right)$$

for $i, j \in \{1, 2\}$, $j \neq i$, $k \in \{0, 1, \ldots, M^j\}$, and $Y \in \mathcal{F}^i$.

The only unknown parameters in (7.8) are the values of i and k, since obviously the coordinates of the optimal new location are unknown. Therefore, all possible values $i = 1, 2$ and $k = 0, \ldots, M^j$ are tested in Algorithm 7.1 to obtain a candidate set for the set of globally optimal solutions.

Algorithm 7.1 is formulated for the case that the corresponding uncon-strained Weber problems $1/P/\bullet/d/\sum$ satisfy the convex hull property (7.3). It can be easily transferred to the case that only the weak convex hull property is satisfied for $1/P/\bullet/d/\sum$.

Algorithm 7.1. (Solution of $1/P/(\mathcal{B}_L, 2\,passages)/d_{\mathcal{B}_L}/\sum$)

Input: *Location problem* $1/P/\mathcal{B}_L/d_{\mathcal{B}_L}/\sum$ *with* $N = 2$ *passages.*

For $i = 1, 2$ *do:*

Step 1:   *Let* $j \in \{1, 2\}$ *with* $j \neq i$.
   Let $D^j(m) := d\left(Ex_m^j, P_1\right) - d\left(Ex_m^j, P_2\right); m \in \mathcal{M}^j$.

Step 2:   *Sort the existing facilities such that* $D^j(1) \leq \cdots \leq D^j(M^j)$.

Step 3:   *For* $k = 0$ *to* M^j, *do:*

   (a) *Let* $w(P_1) := \sum_{m=1}^k w_m^j$
   and $w(P_2) := \sum_{m=k+1}^{M^j} w_m^j$.

   (b) *Determine the set of optimal solutions* $\mathcal{X}_k^i$ *of the corre-sponding unconstrained problem* $1/P/\bullet/d/\sum$ *with exist-ing facilities* $\mathcal{E}x := \left\{Ex_1^i, \ldots, Ex_{M^i}^i, P_1, P_2\right\}$ *and weights of* P_1 *and* P_2 *as defined in (a).*

   (c) *For* $X_k^i \in \mathcal{X}_k^i$ *determine* $F_k^i\left(X_k^i\right) := f_k^i\left(X_k^i\right) + g_k^j$.

Output: $\mathcal{X}_{\mathcal{B}}^* = \displaystyle\operatorname*{arg\,min}_{i \in \{1,2\}\,;\, k \in \mathcal{M}^i\,;\, X_k^i \in \mathcal{X}_k^i} F_k^i\left(X_k^i\right)$.

Note that $f_{\mathcal{B}}(X) \leq F_k^i(X)$ holds for all $X \in \mathcal{F}$, since F_k^i is equivalent to a reformulation $f_Y^i + g_Y^j$ of the objective function with some $Y \in \mathcal{F}^i$ (cf. Corollary 7.1), and thus inequality (7.2) can be applied. Moreover, a solution X_k^i may be found during the algorithm for which the current value of k and therefore the current assignment of passages is not optimal; i.e., $f_{\mathcal{B}}\left(X_k^i\right) < F_k^i\left(X_k^i\right)$. However, the optimal assignment yielding a glob-ally optimal solution $X_{\mathcal{B}}^*$ must be used during the solution process so that equality, i.e., $f_{\mathcal{B}}(X_{\mathcal{B}}^*) = F_k^i(X_{\mathcal{B}}^*)$, holds in this case.

The time complexity of Algorithm 7.1 is $O(M \log M + MT)$, where $M := M^1 + M^2$ is the total number of existing facilities and $O(T)$ is the time complexity for solving a corresponding unconstrained problem of type $1/P/\bullet/d/\sum$. In Step 2, the set of existing facilities are sorted, and in Step 3, M unconstrained Weber problems with time complexity $O(T)$ have to be solved.

Algorithm 7.1 significantly improves the straightforward approach of enu-merating all possible assignments of existing facilities to passage points as described on page 108, which has, in the case of $N = 2$ passage points, an exponential complexity of $O\left(2^N T\right)$.

7.3   Line Barriers with N Passages, $N > 2$

Algorithm 7.1 can be generalized to the case that an arbitrary but finite number N of passages is given that allow trespassing through the line barrier $\mathcal{B}_L$.

Without loss of generality we assume that the passages are given in consecutive order; i.e., there is no passage between P_i and P_{i+1} for $1 \leq i \leq N - 1$. As in Section 7.2, the differences of distances between the existing facilities and every pair of two adjacent passages P_n and P_{n+1} are defined: For each half-plane $\mathcal{F}^i$, $i \in \{1, 2\}$, and for $n = 1, \ldots, N - 1$ let

$$D_n^i(m) := d\left(Ex_m^i, P_n\right) - d\left(Ex_m^i, P_{n+1}\right). \qquad m \in \mathcal{M}^i.$$

Lemma 7.1 implies that a d-shortest permitted path SP between an existing facility $Ex_m^j \in \mathcal{F}^j$, $j \in \{1, 2\}$, $i \neq j$, in the opposite half-plane and a point $X \in \mathcal{F}^i$ has to pass through exactly one of the passages $P_1, \ldots, P_N$ depending on the following condition:

$$
\begin{aligned}
P_1 \in SP \ &\Leftarrow \qquad\qquad\qquad D_1^j(m) \ < \ d(P_2, X) - d(P_1, X), \\
P_n \in SP \ &\Leftarrow \quad d(P_n, X) - d(P_{n-1}, X) \ < \ D_{n-1}^j(m) \\
&\qquad\qquad\qquad\wedge \qquad D_n^j(m) \ < \ d(P_{n+1}, X) - d(P_n, X), \qquad (7.10) \\
P_N \in SP \ &\Leftarrow \quad d(P_N, X) - d(P_{N-1}, X) \ < \ D_{N-1}^j(m).
\end{aligned}
$$

Condition (7.10) must be satisfied for all $X \in \mathcal{F}^i$, since d is a metric induced by a norm and thus $d\left(Ex_m^j, P\right)$ and $d(X, P)$ are convex functions of a passage $P \in L$, moving on the line L, for all $Ex_m^j \in \mathcal{F}^j$. Therefore, either the problem

$$
\begin{aligned}
\min \quad &d\left(Ex_m^j, P\right) + d(P, X) \\
\text{s.t.} \quad &P \in \{P_1, \ldots, P_N\}
\end{aligned}
$$

has a unique minimum, or two or more adjacent passages achieve the same minimum value.

In order to rewrite the objective function for a point $X \in \mathcal{F}^i$ analogously to (7.8), some additional notation will be useful:

For $n = 1, \ldots, N - 1$ let $\pi_n^j : \mathcal{M}^j \to \mathcal{M}^j$ be a permutation of $\mathcal{M}^j$ such that

$$D_n^j\left(\pi_n^j(1)\right) \leq \cdots \leq D_n^j\left(\pi_n^j(M^j)\right).$$

Unfortunately, two permutations π_n^j and $\pi_{\tilde{n}}^j$ need not be the same for $n \neq \tilde{n}$, as can be seen in Figure 7.5.

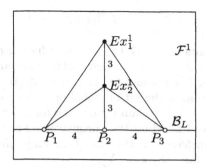

Figure 7.5. An example where $D_1^1(1) = \sqrt{52} - 6 \approx 1.21$ and $D_1^1(2) = 2$, i.e., $D_1^1(1) < D_1^1(2)$, but $D_2^1(1) = -(\sqrt{52} - 6) \approx -1.21$ and $D_2^1(2) = -2$ and thus $D_2^1(2) < D_2^1(1)$.

Furthermore, define $k_N := M^j$, and for $n = 1, \ldots, N - 1$ let

$$k_n := \max\left\{0, \underset{m \in \mathcal{M}^j}{\operatorname{argmax}}\left\{\pi_n^j(m) \; : \; D_n^j\left(\pi_n^j(m)\right) < d(P_{n+1}, X) - d(P_n, X)\right\}\right\}.$$

Using this notation, (7.10) implies that all existing facilities with index $\pi_1^j(m) \in \mathcal{M}^j =: \mathcal{M}_1^j$ and $m \leq k_1$ are assigned to passage P_1. From the remaining existing facilities, i.e., from the existing facilities with index $\pi_2^j(m) \in \mathcal{M}_2^j := \mathcal{M}_1^j \setminus \left\{\pi_1^j : m \leq k_1\right\}$, all those satisfying $m \leq k_2$ are assigned to passage P_2. Since the same argument can be repeated until all existing facilities are assigned to some passage point in $\{P_1, \ldots, P_N\}$, we obtain a total of $\binom{M^j + N - 1}{N - 1}$ feasible assignments of existing facilities in the half-plane $\mathcal{F}^j$ to the passage points $\{P_1, \ldots, P_N\}$.

To formalize the above discussion, we define $\mathcal{M}_1^j := \mathcal{M}^j$ and

$$\mathcal{M}_n^j := \mathcal{M}_{n-1}^j \setminus \left\{\pi_{n-1}^j(m) \; : \; m \leq k_{n-1}\right\}, \quad n = 2, \ldots, N.$$

Then

$$f_\mathcal{B}(X) = f_{k_1,\ldots,k_N}^i(X) + g_{k_1,\ldots,k_N}^j, \tag{7.11}$$

where

$$f_{k_1,\ldots,k_N}^i(Y) \; := \; \sum_{m=1}^{M^i} w_m^i d\left(Y, Ex_m^i\right)$$

$$+ \sum_{n=1}^{N}\left(\sum_{\substack{\pi_n^j(m) \in \mathcal{M}_n^j \\ m \leq k_n}} w_{\pi_n^j(m)}^j\right) d(Y, P_n), \tag{7.12}$$

$$g_{k_1,\ldots,k_N}^j \; := \; \sum_{n=1}^{N} \sum_{\substack{\pi_n^j(m) \in \mathcal{M}_n^j \\ m \leq k_n}} w_{\pi_n^j(m)}^j d\left(P_n, Ex_{\pi_n^j(m)}^j\right),$$

for $i, j \in \{1, 2\}$, $i \neq j$, $0 \leq k_n \leq M^j$ for all $n = 1, \ldots, N - 1$, $k_N = M^j$, and $Y \in \mathcal{F}^i$.

The unknown parameters in (7.11) are the values of i and of $k_1, \ldots, k_N$, since the coordinates of the optimal new location are unknown. Therefore, all possible values for $i = 1, 2$ and for $k_1, \ldots, k_N \in \{0, \ldots, M^j\}$ are tested in Algorithm 7.2 to obtain the set of global optima of $1/P/\mathcal{B}_L/d_{\mathcal{B}_L}/\sum$.

Note that Algorithm 7.2 is again formulated for the case that the convex hull property (7.3) is satisfied for the unconstrained problem $1/P/\bullet/d/\sum$. It can be easily transferred to the case that only the weak convex hull property (7.4) holds for $1/P/\bullet/d/\sum$.

Algorithm 7.2. (Solution of $1/P/(\mathcal{B}_L, N > 2 \text{ passages})/d_{\mathcal{B}_L}/\sum$)

Input: *Location problem* $1/P/\mathcal{B}_L/d_{\mathcal{B}_L}/\sum$ *with* $N > 2$ *passages.*

For $i = 1, 2$ do:

Step 1: *Let $j \in \{1, 2\}$ with $j \neq i$ and*
$$D_n^j(m) := d\left(P_n, Ex_m^j\right) - d\left(P_{n+1}, Ex_m^j\right); \; m \in \mathcal{M}^j;$$
$n = 1, \ldots, N - 1.$

Step 2: *For $n = 1$ to $N - 1$ find a permutation $\pi_n^i : \mathcal{M}^i \to \mathcal{M}^i$ such that $D_n^j\left(\pi_n^j(1)\right) \leq \cdots \leq D_n^j\left(\pi_n^j(M^j)\right)$.*

Step 3: *Let $\mathcal{M}_1^j := \mathcal{M}^j$.*
For $k_1 = 0$ to M^j, do:
Determine $\mathcal{M}_2^j$.
For $k_2 = 0$ to M^j, do:
Determine $\mathcal{M}_3^j$.
$\cdots$

For $k_{N-1} = 0$ to M^j, do:
Determine $\mathcal{M}_N^j$ and let $k_N := M^j$.

(a) *For $n = 1$ to N let*
$$w(P_n) := \sum_{\substack{\pi_n^j(m) \in \mathcal{M}_n^j \\ m \leq k_n}} w_{\pi_n^j(m)}^j.$$

(b) *Determine the set of optimal solutions $\mathcal{X}_{k_1, \ldots, k_N}^i$ of the corresponding unconstrained problem $1/P/\bullet/d/\sum$ with existing facilities $\mathcal{E}x = \left\{Ex_1^i, \ldots, Ex_{M^i}^i, P_1, \ldots, P_N\right\}$ and weights of $P_1, \ldots, P_N$ as defined in (a).*

(c) *For $X_{k_1, \ldots, k_N}^i \in \mathcal{X}_{k_1, \ldots, k_N}^i$ determine*
$$F_{k_1, \ldots, k_N}^i\left(X_{k_1, \ldots, k_N}^i\right) := f_{k_1, \ldots, k_N}^i\left(X_{k_1, \ldots, k_N}^i\right) + g_{k_1, \ldots, k_N}^j.$$

Output: $\mathcal{X}_{\mathcal{B}_L}^* = \arg\min F_{k_1, \ldots, k_N}^i\left(X_{k_1, \ldots, k_N}^i\right).$

With $M := M^1 + M^2$, the time complexity of Algorithm 7.2 can be bounded by $O\left(N(M\log M) + M^{N-1} + \binom{M+N-1}{N-1}T\right)$, where $O(T)$ is the time complexity for solving the unconstrained problems $1/P/\bullet/d/\sum$. Recall that even though $O\left(M^{N-1}\right)$ different vectors $(k_1,\ldots,k_{N-1})$ are tested in Algorithm 7.2, they yield maximally $O\left(\binom{M+N-1}{N-1}\right)$ different assignments of existing facilities to passage points, since all facilities that have been assigned to a passage P_o with $o < n$ are not longer available in the set $\mathcal{M}_n^j$ for an assignment to passage P_n, $n \in \{1,\ldots,N\}$, $j \in \{1,2\}$. To illustrate this, consider the case that all permutations $\pi_1,\ldots,\pi_{N-1}$ are identical. Then only those combinations of $k_1,\ldots,k_{N-1}$ have to be considered that satisfy $0 \le k_1 \le k_2 \le \cdots \le k_N = M^j$. Thus Algorithm 7.2 can be implemented such that it is polynomial with respect to the number of existing facilities if we assume that the number of passage points is fixed and finite. For comparison, the time complexity of the straightforward approach enumerating all feasible assignments of existing facilities to passages (cf. page 108) is $O\left(N^M T\right)$ in the case that N passages are available. Unfortunately, the complexity of Algorithm 7.2 still grows exponentially with respect to the number of passages N.

7.4   Example

To clarify the concepts described in the previous sections, an example with the classification $1/P/(\mathcal{B}_L, 2\ passages)/l_{2,\mathcal{B}_L}/\sum$ will be discussed.

Existing facility Ex_m^i	$w_m^i = w\left(Ex_m^i\right)$	D_m^i
Ex_1^1	$(5,7)^T$	1	-2.24
Ex_2^1	$(4.5,9)^T$	2	-1.99
Ex_3^1	$(10,7.5)^T$	2	3.81
Ex_1^2	$(3,3)^T$	2	-4.09
Ex_2^2	$(6,1)^T$	3	-0.53
Ex_3^2	$(8.5,4)^T$	2	3.49

Table 7.1. Existing facilities with their weights and the values of D_m^i.

In this example a line barrier

$$\mathcal{B}_L := \left\{(x,y)^T \in \mathbb{R}^2 \ : \ y = 5\right\} \setminus \left\{P_1 = (4,5)^T, P_2 = (9,5)^T\right\}$$

divides the plane into two half-planes. Furthermore, three existing facilities are given on either side of $\mathcal{B}_L$ with coordinates and weights as listed in Table 7.1. Thus $\mathcal{M}^1 = \mathcal{M}^2 = \{1, 2, 3\}$ and $M^1 = M^2 = 3$.

Subproblem	Weights		Optimal solutions of the subproblems			
(i, k)	$w(P_1)$	$w(P_2)$	X_k^i	$f_k^i\left(X_k^i\right)$	g_k^j	$F_k^i\left(X_k^i\right)$
$(1, 0)$	0	7	$(9.5)^T$	21.90	29.89	51.79
$(1, 1)$	2	5	$(8.86, 5.14)^T$	31.89	21.71	53.60
$(1, 2)$	5	2	$(4.85, 5.74)^T$	32.78	20.13	52.91
$(1, 3)$	7	0	$(4.5)^T$	23.30	27.11	50.41
$(2, 0)$	0	5	$(8.44, 4.07)^T$	28.47	21.90	50.37
$(2, 1)$	1	4	$(8.14, 4.06)^T$	31.77	19.67	51.44
$(2, 2)$	3	2	$(5.68, 3.43)^T$	32.78	15.69	48.47
$(2, 3)$	5	0	$(4.21, 4.56)^T$	26.99	23.30	50.29

Table 7.2. Optimal solutions of the eight unconstrained subproblems.

Table 7.2 lists the information about the eight unconstrained Weber problems of type $1/P/ \bullet /l_2/ \sum$ that are solved if Algorithm 7.1 is applied.

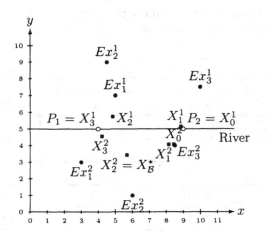

Figure 7.6. The example problem with the classification $1/P/(\mathcal{B}_L, 2\ passages)/l_{2.\mathcal{B}_L}/ \sum$.

The approximate solution values X_k^i and $f_k^i\left(X_k^i\right)$ given in Table 7.2 were obtained using an implementation of Weiszfeld's algorithm

(Weiszfeld, 1937) in LoLA, the Library of Location Algorithms (Hamacher et al., 1999a).

It is easy to see that $X_{\mathcal{B}}^* = (5.72, 3.43)^T$ is a global optimal solution with $f_{\mathcal{B}}(X_{\mathcal{B}}^*) = F_2^2(X_{\mathcal{B}}^*) = 48.47$. In comparison to this result the optimal solution X^* of the corresponding unconstrained Weber problem disregarding the barrier $\mathcal{B}_L$ is given by $X^* = (6.41, 4.40)^T$ with objective function value $f(X^*) = 44.31$.

Part III

Solution Methods for Special Distance and Objective Functions

8

Weber Problems with Block Norms

In this section we develop results for planar location problems that seem to be both of theoretical and practical importance: Block norms (i.e., polyhedral norms, or symmetric polyhedral gauges) are used to evaluate distances, and barriers are introduced that restrict the available area for locating facilities and cannot be crossed in going from one facility to some other. The results discussed in this chapter were also published in Hamacher and Klamroth (2000).

Gauge distances were introduced by Minkowski (1911); see Definition 1.5 on page 6. Within location theory Durier and Michelot (1985) proved a discretization result for location problems with polyhedral gauges, which will be reviewed in the following section. Nickel (1998) showed that location problems with restrictions, i.e., with regions that can be crossed but cannot be used for placement of new facilities, can also be discretized. The importance of polyhedral gauges and, in particular, block norms in evaluating distances in real-world contexts was pointed out by Ward and Wendell (1985) and Brimberg and Love (1995).

Throughout this chapter we assume that $\{B_1, \ldots, B_N\}$ is a finite set of convex, closed, polyhedral and pairwise disjoint barriers in the plane $\mathbb{R}^2$. We use $\mathcal{B} = \bigcup_{i=1}^{N} B_i$ to denote the union of these barrier sets, and $\mathcal{F} = \mathbb{R}^2 \setminus \operatorname{int}(\mathcal{B})$ to denote the feasible region. As before, we assume that the number of extreme points of the barrier polyhedra is finite, and the sets of extreme points and facets of $\mathcal{B}$ are denoted by $\mathcal{P}(\mathcal{B})$ and $\mathcal{F}(\mathcal{B})$, respectively.

Boundedness of the barrier regions is not required in this chapter. Unbounded barriers may occur, for example, in the modeling of oceans and rivers, or, on a smaller scale, in the case that a large barrier region intersects a small modeling horizon.

Furthermore, a finite set of existing facilities $\mathcal{E}x = \{Ex_m \in \mathcal{F} : m \in \mathcal{M} = \{1, \ldots, M\}\}$ is given in a connected subset of the feasible region $\mathcal{F}$. With each existing facility $Ex_m \in \mathcal{E}x$ a positive weight $w_m := w(Ex_m)$ is associated representing the demand of facility Ex_m.

Using this problem formulation, the Weber problem with convex, polyhedral barriers $1/P/(\mathcal{B} = N \, convex \, polyhedra)/d_\mathcal{B}/\sum$ is to find a new facility $X_\mathcal{B}^* \in \mathcal{F}$ such that

$$f_\mathcal{B}(X) = \sum_{i=1}^{M} w_m d_\mathcal{B}(X, Ex_m)$$

is minimized; cf. Section 3.2.

As already mentioned, our main purpose will be to develop concepts for the case that distances are measured by block norms. A block norm γ (polyhedral norm, symmetric polyhedral gauge) according to Definition 1.6 is given by a symmetric convex polyhedron P in the plane $\mathbb{R}^2$ containing the origin $\underline{0} = (0,0)^T$ in its interior. We let $\text{ext}(P) = \{v^1, \ldots, v^\delta\}$ denote the fundamental vectors of P, while $d^1, \ldots, d^\delta$ denote the corresponding fundamental directions; see Figure 8.1.

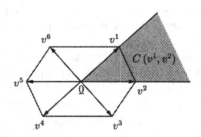

Figure 8.1. A block norm with six fundamental vectors, cf. Figure 2.7.

The methods developed in Sections 4.1 and 4.2 are particularly well suited for finding good bounds for the optimal objective value of Weber problems with barriers and block norms $1/P/\mathcal{B}/\gamma_\mathcal{B}/\sum$ for the following reasons: Recall that a lower bound as well as an upper bound for the optimal objective value of a problem of type $1/P/\mathcal{B}/\gamma_\mathcal{B}/\sum$ can be found by solving the corresponding restricted location problem $1/P/(\mathcal{R} = \mathcal{B})/\gamma/\sum$ with the forbidden region $\mathcal{R} := \mathcal{B}$; see Theorem 4.2 and Corollary 4.2.

Restricted location problems $1/P/\mathcal{R}/\gamma/\sum$ with a Weber objective function and distances measured by block norms (or, more generally, by polyhedral gauges) can be solved by an algorithm developed in Hamacher and Nickel (1994, 1995) for the special case of $\gamma = l_1$ and $\gamma = l_\infty$, and in Nickel

(1995) for general polyhedral gauges. An optimal location $X_{\mathcal{R}}^*$ of the restricted problem can be obtained by first solving the unconstrained problem $1/P/ \bullet /d/ \sum$. If an optimal location X^* of the unconstrained problem is feasible, i.e., $X^* \not\subseteq \text{int}(\mathcal{R})$, then $X_{\mathcal{R}}^* = X^*$ (see Figure 8.2, $\mathcal{B} = \{B_a\}$). Otherwise, it can be shown that $X_{\mathcal{R}}^*$ is the best of the at most δM intersection points of fundamental directions with the boundary $\partial \mathcal{B}$ of $\mathcal{B}$ (see Figure 8.2, $\mathcal{B} = \{B_b\}$).

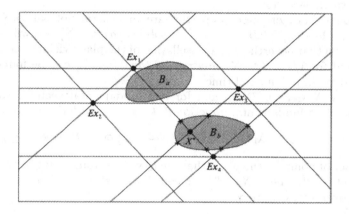

Figure 8.2. In $1/P/(\mathcal{R} = \{B_a\})/\gamma/ \sum$ we have $X_{\mathcal{R}}^* = X^*$. In $1/P/$ $(\mathcal{R} = \{B_b\})/\gamma/ \sum$ one of the intersection points marked by stars is an optimal solution $X_{\mathcal{R}}^*$.

In the example given in Figure 8.2 optimality cannot be shown for the problem $1/P/(\mathcal{R} = \{B_a\})/\gamma/ \sum$ using Corollary 4.1, since $\gamma(Ex_1, X^*) < \gamma_{B_a}(Ex_1, X^*)$.

In the following section we will show that Weber problems with barriers and block norms have some further useful properties. In particular, we will prove that they can be reduced to discrete location problems, i.e., to location problems with a discrete set of possible locations. In Section 8.2 possible generalizations of this result to Weber problems with barriers and general polyhedral gauges are discussed.

8.1 Constructing a Finite Dominating Set

Discretization of planar location problems with polyhedral gauges, a special case of which are block norms, to discrete location problems has already been successful for various kinds of problems. Durier and Michelot (1985) showed that in the case of the unconstrained Weber problem with polyhedral gauges $1/P/ \bullet /\gamma/ \sum$ the fundamental directions rooted at the existing facilities Ex_m, $m \in \mathcal{M}$, the so-called *construction lines*, define a grid tessellation of the plane such that the set of optimal locations is a cell, a line connecting two adjacent grid points of a cell, or a single grid point. If none of these optimal locations is feasible for the restricted Weber problem with convex forbidden regions and polyhedral gauges $1/P/\mathcal{R}/\gamma/\sum$, then Nickel (1995) showed that it is sufficient to consider only the intersection points of construction lines and the boundary $\partial\mathcal{R}$ of the forbidden set $\mathcal{R}$. Both results are heavily based on the fact that the objective function is convex and linear in each cell.

Although both of these properties are in general not satisfied in the barrier problem $1/P/(\mathcal{B} = N$ *convex polyhedra*$)/\gamma_\mathcal{B}/\sum$, we will show in this section that nevertheless, a tessellation of the plane yielding an optimal grid point for $1/P/(\mathcal{B} = N$ *convex polyhedra*$)/\gamma_\mathcal{B}/\sum$ can be found. This can be done in polynomial time.

For any $X \in \mathcal{E}x \cup \mathcal{P}(\mathcal{B})$, and for any fundamental vector v^i and the corresponding fundamental direction d^i ($i = 1, \ldots, \delta$) let

$$\left(X + d^i\right)_\mathcal{B} := \left\{X + \lambda v^i \ : \ \lambda \in \mathbb{R}_+; \ (X + \mu v^i) \cap \text{int}(\mathcal{B}) = \emptyset \ \forall 0 \leq \mu \leq \lambda\right\}$$

be the set of points in the plane that are l_2-visible from X in the fundamental direction d^i. Then $\left(X + d^i\right)_\mathcal{B}$ is called a *construction line* with respect to the given block norm γ.

Definition 8.1. *The grid*

$$\mathcal{N}_\gamma := \left(\bigcup_{X \in \mathcal{E}x \cup \mathcal{P}(\mathcal{B})} \bigcup_{i=1}^{\delta} \left(X + d^i\right)_\mathcal{B}\right) \cup \mathcal{F}(\mathcal{B})$$

is called a construction line grid *with respect to $\mathcal{E}x$, $\mathcal{B}$, and the block norm γ.*

The intersection points of line segments in $\mathcal{N}_\gamma$ define the set $\mathcal{P}(\mathcal{N}_\gamma)$ of grid points, and $\mathcal{C}(\mathcal{N}_\gamma)$ is the set of resulting cells in $\mathcal{F}$, i.e., the set of smallest convex polyhedra with nonempty interior and with extreme points in $\mathcal{P}(\mathcal{N}_\gamma)$. An example for the construction of a grid $\mathcal{N}_\gamma$ is given in Figure 8.3. Note that all cells $C \in \mathcal{C}(\mathcal{N}_\gamma)$ are convex polyhedral sets, since they are bounded by construction lines rooted at extreme points of barriers or at existing facilities, and by facets of the convex barrier polyhedra.

The grid $\mathcal{N}_\gamma$ is constructed such that each existing facility in $\mathcal{E}x$ and each extreme point in $\mathcal{P}(\mathcal{B})$ of a barrier that is γ-visible from one point in the interior of a cell C is γ-visible from all points in C. In other words, no cell in $\mathcal{C}(\mathcal{N}_\gamma)$ is intersected by the boundary of the γ-shadow of an existing facility or of an extreme point of a barrier. Moreover, the grid $\mathcal{G}_\gamma$ as defined in Definition 5.2 on page 60 is contained in the grid $\mathcal{N}_\gamma$ as a subgrid; i.e., $\mathcal{G}_\gamma \subseteq \mathcal{N}_\gamma$. Thus the following corollary is an immediate consequence of Lemma 5.1 on page 62.

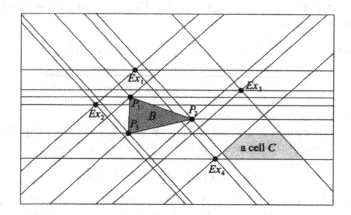

Figure 8.3. The grid $\mathcal{N}_\gamma$ for a barrier problem with the block norm introduced in Figure 8.1.

Corollary 8.1. *Let $I \in \mathcal{E}x \cup \mathcal{P}(\mathcal{B})$ be an existing facility or an extreme point of a barrier, and let C be a cell in $\mathcal{C}(\mathcal{N}_\gamma)$. If I is γ-visible from one point in $\mathrm{int}(C)$, then I is γ-visible from all points in C.*

Similarly, the following corollary is an immediate consequence of Lemma 5.2 on page 63, since $\mathcal{G}_\gamma \subseteq \mathcal{N}_\gamma$.

Corollary 8.2. *Let $C \in \mathcal{C}(\mathcal{N}_\gamma)$ be a cell, let $X \in C$ be a feasible solution of $1/P/(\mathcal{B} = N$ convex polyhedra$)/\gamma_\mathcal{B}/\sum$, and let $Ex \in \mathcal{E}x$. Then there exists an intermediate point $I = I_{Ex,X}$ on a γ-shortest permitted Ex-X path that is γ-visible from all points in C.*

The following result proves the existence of a finite dominating set for problems of the type $1/P/(\mathcal{B} = N$ convex polyhedra$)/\gamma_\mathcal{B}/\sum$, i.e., a finite set of points containing at least one optimal solution of the problem.

Theorem 8.1. *One of the grid points of $\mathcal{N}_\gamma$ is optimal for $1/P/(\mathcal{B} = N$ convex polyhedra$)/\gamma_\mathcal{B}/\sum$.*

Proof. Let $C \in \mathcal{C}(\mathcal{N}_\gamma)$ be a cell and let $X \in C$ be such that X is not a grid point. Moreover, for $m = 1, \ldots, M$ let $I_m \in \mathcal{E}x \cup \mathcal{P}(\mathcal{B})$ be an intermediate point on a γ-shortest permitted Ex_m-X path according to Definition 2.5 and such that I_m is γ-visible from all points in C (see Figure 8.4). Note that the existence of intermediate points with this property follows from Corollary 8.2.

Since $\gamma_{\mathcal{B}}(X, Ex_m) = \gamma(X, I_m) + \gamma_{\mathcal{B}}(I_m, Ex_m)$, the objective function for X can be written as

$$f_{\mathcal{B}}(X) = \underbrace{\sum_{m \in \mathcal{M}} w_m \gamma(X, I_m)}_{=f_X(X)} + \underbrace{\sum_{m \in \mathcal{M}} w_m \gamma_{\mathcal{B}}(I_m, Ex_m)}_{=g_X \text{ (constant for fixed } X)}; \tag{8.1}$$

compare also (5.3) on page 77.

For any other point $Y \in C$ we have $\gamma_{\mathcal{B}}(Y, Ex_m) \le \gamma(Y, I_m) + \gamma_{\mathcal{B}}(I_m, Ex_m)$, since I_m is γ-visible from every point of the cell C and thus

$$f_{\mathcal{B}}(Y) \le f_X(Y) + g_X \qquad \forall Y \in C.$$

where equality holds for $X = Y$. Here $f_X(Y)$ is the objective function of an unconstrained Weber problem $1/P/ \bullet /\gamma/ \sum$ with existing facilities $\{I_m : m \in \mathcal{M}\}$.

Ward and Wendell (1985) proved for this problem $1/P/ \bullet /\gamma/ \sum$ that the level curves $L_=(z, f_X, C) := \{Y \in C : f_X(Y) = z\}$ are linear in the cell C. Furthermore, note that the cell C of the grid $\mathcal{N}_\gamma$ is contained in a cell $\overline{C}$ of the analogous grid $\overline{\mathcal{N}_\gamma}$ of this unconstrained Weber problem $1/P/ \bullet /\gamma/ \sum$. From the convexity of C it follows that there must exist a grid point $I^* \in \mathcal{P}(\mathcal{N}_\gamma)$ of C such that $f_X(I^*) \le f_X(X)$. Hence

$$f_{\mathcal{B}}(I^*) \le f_X(I^*) + g_X \le f_X(X) + g_X = f_{\mathcal{B}}(X),$$

proving Theorem 8.1. □

It should be noted that this result has long been known for rectilinear distances ($\gamma = l_1$) (see Larson and Sadiq, 1983). Their proof relies heavily on the fact that the objective function is convex within each cell, a fact that is not needed in the preceding proof. Moreover, Larson and Sadiq (1983) proved in the rectilinear case that for any point X in a cell C there exists an l_1-shortest permitted path from X to Ex_m passing through a corner point of C. This is in general not true for block norms. In Figure 8.4 there exists, for example, no γ-shortest permitted path from X to Ex_3 passing through a corner point of the shaded cell C.

The methods used in the proof of Theorem 8.1 will be generalized in the following to derive a stronger result for the complete set of optimal solutions of $1/P/(\mathcal{B} = N \text{ convex polyhedra})/\gamma_{\mathcal{B}}/ \sum$.

Theorem 8.2. *The set $\mathcal{X}_{\mathcal{B}}^*$ of optimal solutions of $1/P/(\mathcal{B} = N \text{ convex polyhedra})/\gamma_{\mathcal{B}}/ \sum$ can be decomposed into subsets that are either*

- *grid points of $\mathcal{N}_\gamma$,*

- *facets of cells of $\mathcal{N}_\gamma$ or*

- *complete cells of $\mathcal{N}_\gamma$.*

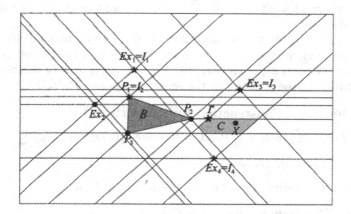

Figure 8.4. For a point $X \in C$ the corresponding intermediate points I_m, $m = 1, \ldots, 4$, and I^* are marked by stars.

Proof. Let $C \in \mathcal{C}(\mathcal{N}_\gamma)$ be a cell and let $X \in int(C)$ be a point in $\mathcal{X}_{\mathcal{B}}^*$ with optimal objective value $f_{\mathcal{B}}(X) = z^*$.

Using the same decomposition of the objective function for the point X as in the proof of Theorem 8.1, i.e., $f_{\mathcal{B}}(X) = f_X(X) + g_X = z^*$, $f_X(X) = z_X^*$ is the minimal objective value among all points in the cell C of the unconstrained Weber problem $1/P/ \bullet /\gamma/ \sum$ with respect to the intermediate points I_m (see the proof of Theorem 8.1). Since Ward and Wendell (1985) proved the linearity of the level curves for this problem within every cell C, the level curve $L_=(z_X^*, f_X, C)$ passing through X must contain the complete cell C; i.e., $f_X(Y) = z_X^*$ holds for all points $Y \in C$. Thus, using $f_{\mathcal{B}}(Y) \leq f_X(Y) + g_X = z_X^* + g_X$, we obtain that $f_{\mathcal{B}}(Y) = z^*$ for all points $Y \in C$; i.e., the complete cell C is optimal for $1/P/(\mathcal{B} = N \text{ convex polyhedra})/\gamma_{\mathcal{B}}/ \sum$.

The same argument can be used to show that if a point X on a facet of a cell (that is not a corner point) is optimal for $1/P/(\mathcal{B} = N \text{ convex polyhedra})/\gamma_{\mathcal{B}}/ \sum$, the complete facet must be optimal, completing the proof of Theorem 8.2. $\qquad \square$

Note that Theorems 8.1 and 8.2 differ from Theorem 5.5 on page 78 in the sense that they do not reduce the barrier problem $1/P/(\mathcal{B} = N \text{ convex polyhedra})/\gamma_{\mathcal{B}}/ \sum$ to a finite set of unconstrained subproblems $1/P/ \bullet /\gamma/ \sum$ within each cell of the subgrid $\mathcal{G}_\gamma$ of $\mathcal{N}_\gamma$, but that they use a further decomposition of the cells in $\mathcal{G}_\gamma$ to construct a finite dominating set for $1/P/(\mathcal{B} = N \text{ convex polyhedra})/\gamma_{\mathcal{B}}/ \sum$. This makes the consideration of all feasible assignments of intermediate points to existing facilities obsolete (cf. Sections 5.3 and 5.5). We will see later in this section that the constructed finite dominating set is in fact of polynomial cardinality, which significantly improves the results of Section 5.5 for the special case of block norm distances.

Theorem 8.1 leads to the formulation of a simple and efficient algorithm that computes at least one optimal solution of the barrier problem $1/P/(\mathcal{B} = N\ convex\ polyhedra)/\gamma_\mathcal{B}/\sum$. The algorithm is based on the discretization of the problem to the set of grid points $\mathcal{P}(\mathcal{N}_\gamma)$.

Algorithm 8.1. (Construction Line Algorithm)

Input: *Location problem* $1/P/(\mathcal{B} = N\ convex\ polyhedra)/\gamma_\mathcal{B}/\sum$.

Step 1:   *Compute the grid* $\mathcal{N}_\gamma$.

Step 2:   *Determine the set of all grid points* $\mathcal{P}(\mathcal{N}_\gamma)$.

Output: $X_\mathcal{B}^* \in \operatorname{argmin}\{f_\mathcal{B}(I) : \ I \in \mathcal{P}(\mathcal{N}_\gamma)\}$.

The size of the finite dominating set $\mathcal{P}(\mathcal{N}_\gamma)$ constructed during this algorithm is bounded by the number of intersection points of construction lines in $\mathcal{N}_\gamma$. Since the total number of construction lines is bounded by $O((|\mathcal{E}x| + |\mathcal{P}(\mathcal{B})|)\delta)$ (recall that δ denotes the number of fundamental vectors of the given block norm γ) and since each construction line may intersect every other construction line at most once, the size of the candidate set $|\mathcal{P}(\mathcal{N}_\gamma)|$ is bounded by $O\left((|\mathcal{E}x| + |\mathcal{P}(\mathcal{B})|)^2\delta^2\right)$. The construction line algorithm thus solves problems of the type $1/P/(\mathcal{B} = N\ convex\ polyhedra)/\gamma_\mathcal{B}/\sum$ in polynomial time.

In comparing this result to the unconstrained case, i.e., $|\mathcal{P}(\mathcal{B})| = 0$, we see that the construction line algorithm developed in Durier and Michelot (1985) utilizes a grid based on $O(|\mathcal{E}x|\delta)$ construction lines with at most $O\left(|\mathcal{E}x|^2\delta^2\right)$ intersection points. Thus Algorithm 8.1 can be seen as a natural extension of the unconstrained case, where the number (and complexity in terms of extreme points) of the introduced barrier regions has a similar impact on the complexity of the problem as the number of existing facilities.

It will be shown in the following that the size of the finite dominating set for $1/P/(\mathcal{B} = N\ convex\ polyhedra)/\gamma_\mathcal{B}/\sum$ can be further reduced by omitting a large subset of points that cannot be optimal. The optimal solution will be restricted to the iterative convex hull $H_\mathcal{B}$ of the existing facilities and the barrier regions (cf. Definition 5.4 on page 67).

Theorem 8.3. *Let $H_\mathcal{B}$ be the iterative convex hull of the existing facilities in $\mathcal{E}x$ and the barrier regions in $\mathcal{B}$. Then at least one optimal solution of the barrier problem $1/P/(\mathcal{B} = N\ convex\ polyhedra)/\gamma_\mathcal{B}/\sum$ exists in a grid point in $H_\mathcal{B}$.*

Proof. Let $\mathcal{X}_\mathcal{B}^*$ be the set of optimal locations of $1/P/(\mathcal{B} = N\ convex\ polyhedra)/\gamma_\mathcal{B}/\sum$. Suppose that $\mathcal{X}_\mathcal{B}^* \cap H_\mathcal{B} = \emptyset$ and choose some $X^* \in \mathcal{X}_\mathcal{B}^*$ with $f_\mathcal{B}(X^*) = z^*$. We can assume that there exists no barrier in $\mathbb{R}^2 \setminus H_\mathcal{B}$, since this assumption may only decrease the objective value of X^*, while it has no influence on the objective value of points in $H_\mathcal{B} \cap \mathcal{F}$.

For each existing facility $Ex_m \in \mathcal{E}x$ there exists a γ-shortest permitted path to X^* that intersects the boundary $\partial(H_\mathcal{B})$ of $H_\mathcal{B}$ in a first point I_m such that I_m is l_2-visible from X^* (see Lemma 2.3 on page 31). All these transition points I_m, $m \in \mathcal{M}$, are therefore located on those faces $F^i(H_\mathcal{B})$, $i = 1, \ldots, k$, of the boundary $\partial(H_\mathcal{B})$ of $H_\mathcal{B}$ that are l_2-visible from X^*; see Figure 8.5.

Since $H_\mathcal{B}$ is the convex hull of a set of points and a set of convex polyhedra, $H_\mathcal{B}$ itself is also a convex polyhedron. Furthermore, $X^* \notin H_\mathcal{B}$ and the supporting hyperplanes h^i defining the faces F^i divide $\mathbb{R}^2$ into two half-planes H_1^i and H_2^i such that $X^* \in H_1^i$ and $H_\mathcal{B} \subset H_2^i$ ($i = 1, \ldots, k$). Hence for each existing facility Ex_m ($m \in \mathcal{M}$) the straight line segment connecting X^* and I_m intersects h^i in a point I_m^i; see Figure 8.5.

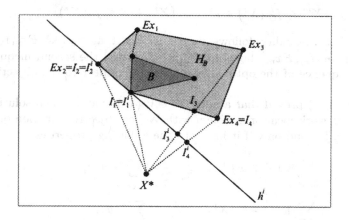

Figure 8.5. The transition points I_m and I_m^i for the point $X^* \notin H_\mathcal{B}$ in the example problem.

The objective function value $f_\mathcal{B}(X^*)$ can therefore be determined in k alternative ways, each one with respect to the transition points I_m^i, $i = 1, \ldots, k$, as

$$f_\mathcal{B}(X^*) = \underbrace{\sum_{m=1}^{M} w_m \gamma\left(X^*, I_m^i\right)}_{=:f^i(X^*)} + \underbrace{\sum_{m=1}^{M} w_m \gamma_\mathcal{B}\left(I_m^i, Ex_m\right)}_{=:\kappa^i}, \qquad i \in \{1, \ldots, k\}.$$

For $i \in \{1, \ldots, k\}$, κ^i is constant for each i, and f^i is the objective function of an unconstrained Weber problem $1/P/\bullet/\gamma/\sum$ with existing facilities I_m^i, $m \in \mathcal{M}$. This problem has at least one optimal solution in $\mathrm{conv}\{I_m^i : m \in \mathcal{M}\}$ (see Durier and Michelot, 1985), namely, on the line h^i.

Now consider the node network location problem $1/T^i/\bullet/d(V,V)/\sum$ on the tree T^i defined by the node set $V\left(T^i\right) = \left\{I_m^i : m \in \mathcal{M}\right\}$ and weights $w(v) = w(Ex_m)$ if $v = I_m^i, m \in \mathcal{M}$. Two nodes $v = I_m^i, u = I_n^i \in V\left(T^i\right)$ are connected by an edge of length $d_G(u,v) := \gamma\left(I_m^i, I_n^i\right)$ if the corresponding points I_m^i and I_n^i of the planar embedding of T^i on h^i are consecutive points on h^i. Then a Weber problem on T^i is given by

$$\min \quad f_G(X) = \sum_{v \in V(T^i)} w(v)\, d_G(v, X)$$
$$\text{s.t.} \quad X \in V\left(T^i\right).$$

The optimal solution X^i of this node network location problem is also optimal for the unconstrained Weber problem with objective function f^i, since $f^i(X) = f_G^i(X)$ holds for all X in the embedding of T^i on h^i. Moreover, X^i satisfies

$$f_B\left(X^i\right) \le f_G^i\left(X^i\right) + \kappa^i = f^i\left(X^i\right) + \kappa^i \le f_B(X^*),$$

where the first inequality follows from the fact that $\gamma_B\left(X^i, Ex_m\right) \le \gamma\left(X^i, I_m^i\right) + \gamma_B\left(I_m^i, Ex_m\right)$ holds for all $m \in \mathcal{M}$, and the second inequality is a consequence of the optimality of X^i with respect to the objective function f^i.

Goldman (1971) proved that a node $X^i \in V\left(T^i\right)$ is an optimal solution of the node network location problem with Weber objective function on a tree network T^i if and only if it has both the following properties:

$$\sum_{v \in V^i} w(v) + w\left(X^i\right) \ge \frac{1}{2} \sum_{v \in V(T^i)} w(v),$$

$$\sum_{v \in \bar{V}^i} w(v) + w\left(X^i\right) \ge \frac{1}{2} \sum_{v \in V(T^i)} w(v),$$

where V^i and $\bar{V}^i$ are the two disjoint connected components of $V\left(T^i\right)$ resulting from the removal of node X^i. These two properties depend only on the weights of the nodes and on their order on h^i, which is identical for all $i \in \{1, \ldots, k\}$. Thus there exists an index $m \in \mathcal{M}$ such that $X^i = I_m^i$ is an optimal solution of $1/T^i/\bullet/d(V,V)/\sum$ for all $i \in \{1, \ldots, k\}$ and

$$f_B\left(I_m^i\right) \le f_B(X^*); \qquad i \in \{1, \ldots, k\}.$$

Since the point I_m^i has to be located on the boundary of $\partial(H_B)$ for at least one index $i \in \{1, \ldots, k\}$, this fact contradicts the assumption $\mathcal{X}_B^* \cap H_B = \emptyset$. Using Theorem 8.2, it can be concluded that there exists at least one optimal grid point in H_B. □

Observe that Theorem 8.3 cannot be deduced from Theorem 5.3 on page 69, since the set of optimal solutions of the unconstrained problem $1/P/\bullet$

$/\gamma/\sum$ is in general not completely contained within the convex hull of the existing facilities (see Durier and Michelot, 1985), and thus the assumption of Theorem 5.3 is not satisfied in this case.

The set $H_\mathcal{B}$ of Theorem 8.3 can be determined using Algorithm 5.1 on page 67.

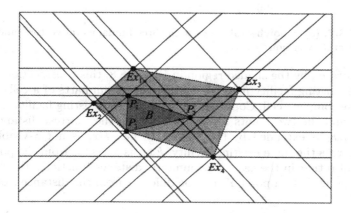

Figure 8.6. Applying Theorem 8.3, the candidate set of the example problem can be reduced from 83 candidate points to only 35 candidate points in the set $H_\mathcal{B}$.

Figure 8.6 indicates the reduced number of points that have to be investigated during the construction line algorithm, Algorithm 8.1, if Theorem 8.3 is applied. However, the theoretical bound on the size of the candidate set $\mathcal{P}(\mathcal{N}_\gamma)$ is not affected by Theorem 8.3, since, for example, in the case of unbounded barriers, a reduction may not be possible.

8.2 Generalization to Polyhedral Gauges

In the previous section a discretization result for Weber problems with barriers and block norms was proven. This result implies a polynomial algorithm to solve problems of type $1/P/(\mathcal{B} = N \ convex \ polyhedra)/\gamma_\mathcal{B}/\sum$.

Even though we mainly focused on block norms, the results can be generalized to the more general class of polyhedral gauges. In Figure 8.7 (a), an example of a polyhedral gauge is given that is not a block norm.

Since polyhedral gauges may be asymmetric, some additional considerations have to be made. In Figure 8.8 the grid as defined in Section 8.1 is shown, using the three fundamental vectors v^1, v^2, v^3 from Figure 8.7 (a) and the corresponding fundamental directions d^1, d^2, d^3.

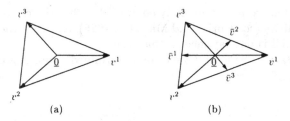

(a)   (b)

Figure 8.7. (a) A polyhedral gauge with three fundamental vectors, and (b) its fundamental extension.

As illustrated by the shaded region in Figure 8.8, this construction implies that an existing facility may be γ-visible only from parts of a cell and not from the complete cell. Consider, for example, the existing facility Ex_1, the cell C, and the two points X_1, X_2 in C. Then the shortest distance from X_2 to Ex_1 is extended by the barrier B; i.e., Ex_1 is not γ-visible from X_2, whereas the same existing facility, Ex_1, is γ-visible from the point X_1. Recall also that in the case of asymmetric distance functions, the distance from a point X to a point Y may be different from the distance from Y to X.

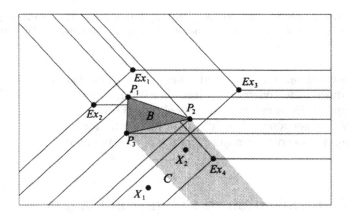

Figure 8.8. The grid $\mathcal{N}_\gamma$ for the example problem with the polyhedral gauge given in Figure 8.7 (a).

In order to overcome this difficulty, which was relevant in the proof of Theorem 8.1, the *fundamental extension* of a given asymmetric polyhedral gauge as shown in Figure 8.7 (b) can be used. The basic idea is to introduce the redundant fundamental vectors $\bar{v}^1, \bar{v}^2$, and $\bar{v}^3$ pointing in the opposite directions to the given fundamental vectors v^1, v^2, and v^3. These additional fundamental vectors do not influence the distance measure with respect to the given polyhedral gauge γ. But their existence implies a finer grid $\overline{\mathcal{N}_\gamma}$ in which every existing facility that is γ-visible from some point in the interior of a cell is also γ-visible from every other point in the same cell.

Based on this grid $\overline{\mathcal{N}_\gamma}$ the results of the previous sections can be generalized to Weber problems with asymmetric polyhedral gauges.

9
Center Problems with the Manhattan Metric

Based on the work Dearing et al. (2002), this chapter considers center problems with barriers.

We consider a finite set of compact, convex, polyhedral, and pairwise disjoint barriers $\{B_1, \ldots, B_N\}$ in the plane $\mathbb{R}^2$, define $\mathcal{B} = \bigcup_{i=1}^{N} B_i$, $\mathcal{F} = \mathbb{R}^2 \setminus \text{int}(\mathcal{B})$, and denote the finite sets of extreme points and facets of the barrier polyhedra by $\mathcal{P}(\mathcal{B})$ and $\mathcal{F}(\mathcal{B})$, respectively. Furthermore, a finite set $\mathcal{E}x = \{Ex_m \in \mathcal{F} : m \in \mathcal{M} = \{1, \ldots, M\}\}$ of existing facilities is given in the feasible region $\mathcal{F}$. A positive weight $w_m = w(Ex_m)$, $m \in \mathcal{M}$, is associated with each existing facility that represents the demand of facility Ex_m. To avoid trivial cases, we assume that $M \geq 3$.

As distance measure the Manhattan metric l_1 will be used, and the corresponding barrier distance $l_{1,\mathcal{B}}$ is defined according to Definition 2.2 on page 17. Thus the center objective function is given by

$$f_{\mathcal{B}}(X) = \max_{m \in \mathcal{M}} w_m l_{1,\mathcal{B}}(X, Ex_m);$$

cf. Section 3.2. The center problem with convex, polyhedral barriers and the Manhattan metric, classified as $1/P/(\mathcal{B} = N\ convex\ polyhedra)/l_{1,\mathcal{B}}/\max$, is to minimize $f_{\mathcal{B}}(X)$ for all $X \in \mathcal{F}$.

Center location problems in the plane without barriers have been extensively discussed in the literature; see, for example, the textbooks of Drezner (1995), Francis et al. (1992), Hamacher (1995), and Love et al. (1988) and the survey of Plastria (2002).

Due to the introduction of the barrier regions, the objective function $f_{\mathcal{B}}$ is nonconvex. Instead of tackling the problem with methods of nonconvex optimization the internal structure of the problem is exploited in more detail. For this purpose, we focus on the Manhattan metric l_1 and the corresponding barrier distance function $l_{1.B}$, i.e., on problems of the type $1/P/(\mathcal{B} = N \, convex \, polyhedra)/l_{1.B}/\max$.

A discretization result for center problems with barriers and the Manhattan metric l_1 is developed in Sections 9.1 to 9.3. It will be shown that there exists a dominating set based on the intersection of weighted bisectors between pairs of existing facilities that contains at least one optimal solution of $1/P/(\mathcal{B} = N \, convex \, polyhedra)/l_{1.B}/\max$. The resulting algorithm using this dominating set is given in Section 9.3. Section 9.4 shows that the results of Sections 9.1 to 9.3 are more general than one might initially think: Any problem with block norms having exactly four fundamental vectors can be tackled by the same approach.

9.1   A Cell Decomposition of the Feasible Region

In the following a network N_{l_1} will be constructed such that l_1-shortest permitted paths between all existing facilities and extreme points of barriers are represented by network paths in N_{l_1}, similar to the visibility graph given in Section 3.3; see Figure 3.1 on page 47. Additional edges are added, resulting in a decomposition of the feasible region into cells, similar to the grid structure used in Section 8.1; cf. Definition 8.1 on page 124. The four fundamental vectors $v^1 = (0.1)^T$, $v^2 = (1.0)^T$, $v^3 = (0,-1)^T$, and $v^4 = (-1,0)^T$, defining the unit ball of the Manhattan metric (cf. Figure 1.5 on page 11), and the corresponding fundamental directions $d^i = \{\lambda v^i : \lambda \geq 0\}$, $i = 1, \ldots, 4$, play a central role in the construction of the network N_{l_1}.

For every $X \in \mathcal{E}x \cup \mathcal{P}(\mathcal{B})$ and for every fundamental vector v^i and the corresponding fundamental direction d^i, $i = 1, \ldots, 4$, a *construction line*

$$(X + d^i)_{\mathcal{B}} := \{X + \lambda v^i \; : \; \lambda \in \mathbb{R}_+ : (X + \mu v^i) \cap \mathrm{int}(\mathcal{B}) = \emptyset \; \forall 0 \leq \mu \leq \lambda\}$$

is defined as the set of points in the plane that are l_2-visible from X in the fundamental direction d^i. Then

$$N_{l_1} := \left(\bigcup_{X \in \mathcal{E}x \cup \mathcal{P}(\mathcal{B})} \bigcup_{i=1}^{4} (X + d^i)_{\mathcal{B}} \right) \cup \mathcal{F}(\mathcal{B}) \tag{9.1}$$

defines a grid in $\mathcal{F}$ according to Definition 8.1. Moreover, a network N_{l_1} corresponding to the grid N_{l_1} is defined as follows: All possible intersection points of construction lines, or the intersection points of construction lines with facets of barriers in N_{l_1} define the set $V(N_{l_1}) := \mathcal{P}(N_{l_1})$ of vertices of the corresponding network N_{l_1}. Two vertices $u_1, u_2 \in V(N_{l_1})$ are connected by an edge in $E(N_{l_1})$ if they are adjacent on some construction line or barrier facet in N_{l_1}. The length of this edge is then given by the l_1-length of the corresponding line segment in N_{l_1}.

The grid N_{l_1} decomposes the feasible region $\mathcal{F}$ into a finite set of cells denoted by $C(N_{l_1})$, i.e., the set of smallest (convex) polyhedra with nonempty interior and with extreme points in $P(N_{l_1})$; see Figure 9.1. The extreme points of a cell $C \in C(N_{l_1})$ are called *corner points* of the cell C. Note that the boundary of each cell consists of construction lines or facets of the barriers.

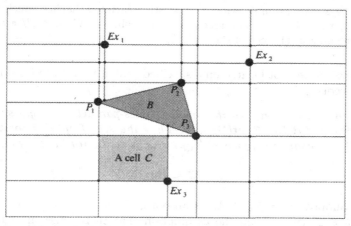

Figure 9.1. The grid N_{l_1} for an example problem with three existing facilities and one triangular barrier. The vertices of the corresponding network N_{l_1} are represented by small dots.

As discussed already on page 126 in Section 8.1, Larson and Sadiq (1983) defined a network similar to the network N_{l_1}, omitting some of the construction lines introduced above. Even though the properties of l_1-shortest permitted paths with respect to their smaller network need some further discussion, Larson and Sadiq showed that the Weber problem with polyhedral barriers $1/P/(\mathcal{B} = N\ polyhedra)/l_{1,\mathcal{B}}/\sum$ can be transformed into a network location problem on the corresponding network. Observe that an analogous result is not true for the center objective function even in the unconstrained case, as can be seen in Figure 9.2.

In analogy to the general results of Section 5.2 it can nevertheless be shown that the set of optimal solutions $\mathcal{X}_{\mathcal{B}}^*$ of $1/P/(\mathcal{B} = N\ convex\ polyhedra)/l_{1,\mathcal{B}}/\max$ is contained in the *iterative rectangular hull* $R_{\mathcal{B}}$ of the existing facilities and the barrier regions, i.e., in the smallest rectangle $R_{\mathcal{B}}$ with sides parallel to the coordinate axes containing all existing facilities and such that the boundary $\partial(R_{\mathcal{B}})$ of $R_{\mathcal{B}}$ does not intersect with the interior of a barrier. However, observe again that the following Theorem 9.1 extends the results of Corollary 5.2 on page 70, since the restriction to

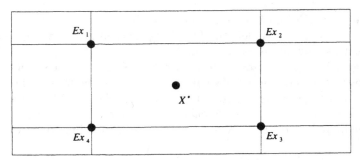

Figure 9.2. An example with four existing facilities having equal weights $w_m = 1$, $m = 1, \ldots, 4$. The unique optimal solution X^* of $1/P/ \bullet /l_1/ \max$ lies in the interior of a cell of the grid $\mathcal{N}_{l_1}$.

optimal solutions not located on a specified grid as utilized in Corollary 5.2 is not needed.

Theorem 9.1. *Let $R_\mathcal{B}$ be the smallest axes-parallel rectangle such that $\mathcal{E}x \subseteq R_\mathcal{B}$ and $\partial(R_\mathcal{B}) \cap \mathrm{int}(\mathcal{B}) = \emptyset$. Then the set of optimal solutions of $1/P/(\mathcal{B} = N \text{ convex polyhedra})/l_{1,\mathcal{B}}/ \max$ is contained in $R_\mathcal{B}$; i.e.,*

$$\mathcal{X}_\mathcal{B}^* \subseteq R_\mathcal{B}.$$

Proof. Suppose that $X_\mathcal{B}^* \in \mathcal{X}_\mathcal{B}^*$ is an optimal solution not located in $R_\mathcal{B}$; i.e., $X_\mathcal{B}^* \in \mathcal{F} \setminus R_\mathcal{B}$. Then the assumption that there exists no barrier in $\mathbb{R}^2 \setminus R_\mathcal{B}$ does not increase the objective value of $X_\mathcal{B}^*$, and it has no influence on the objective values of points in $R_\mathcal{B}$.

Let $R_\mathcal{B} = [x_1, x_2] \times [y_1, y_2]$ and let $X_\mathcal{B}^* = (a, b)^T$. Since $X_\mathcal{B}^* \notin R_\mathcal{B}$, we can conclude that a is not contained in the closed interval $[x_1, x_2]$ or that b is not contained in the closed interval $[y_1, y_2]$; without loss of generality let $a < x_1$.

Let SP_m be an l_1-shortest permitted path from $X_\mathcal{B}^*$ to Ex_m with the barrier touching property of Lemma 2.3 (see page 31), and let $I_m \in \mathcal{P}(\mathcal{B}) \cup \{Ex_m\}$ be an intermediate point on SP_m that is l_2-visible from $X_\mathcal{B}^*$, $m \in \mathcal{M}$. The straight line segment connecting $X_\mathcal{B}^*$ and I_m, $m \in \mathcal{M}$, intersects $\partial(R_\mathcal{B})$ in a point $(a_m, b_m)^T$ with $a < a_m$. Thus moving $X_\mathcal{B}^*$ towards the boundary of $R_\mathcal{B}$ by increasing a to $a + \epsilon$ with a small $\epsilon > 0$ such that $a + \epsilon < x_1$ decreases the distance between $X_\mathcal{B}^*$ and $(a_m, b_m)^T$, and thus also between $X_\mathcal{B}^*$ and each of the intermediate points I_m and between $X_\mathcal{B}^*$ and Ex_m, contradicting the optimality of $X_\mathcal{B}^*$. $\square$

The iterative rectangular hull $R_\mathcal{B}$ can be constructed similarly to the iterative convex hull of Section 5.2; see Algorithm 5.1 on page 67, where in each iteration of the algorithm the determination of the convex hull of the new set of points is replaced by the determination of the smallest axes-parallel rectangle containing all the specified points. Note that since the barrier polyhedra are compact sets, R_B is a compact and convex set the boundary of which is part of the network N_{l_1}.

Since the boundary $\partial(R_B)$ of the iterative rectangular hull R_B is feasible, i.e., $\partial(R_B) \subseteq \mathcal{F}$, the following result for l_1-shortest permitted paths between two points in R_B can be proven:

Lemma 9.1. *Let X and Y be two points in $\mathcal{F} \cap R_B$. Then every l_1-shortest permitted path connecting X and Y lies completely in $\mathcal{F} \cap R_B$.*

Proof. The result follows from the fact that all barriers $B \subseteq \mathbb{R}^2 \setminus R_B$ can be discarded. Every permitted path connecting X and Y that leaves the set $\mathcal{F} \cap R_B$ at some point $Z_1 \in \partial(R_B)$ has to reenter at some point $Z_2 \in \partial(R_B)$ and will be dominated by a path using the shortest path from Z_1 to Z_2 along the boundary of R_B instead. $\square$

Note that Theorem 9.1 implies not only that it is sufficient to consider only cells within the rectangle R_B, but that, using Lemma 9.1, we can also reduce the network N_{l_1} to the subnetwork $N'_{l_1} \subseteq N_{l_1}$ that results from the intersection of the embedding of N_{l_1} in $\mathcal{F}$ with the rectangle R_B.

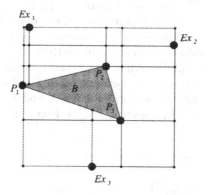

Figure 9.3. The rectangle R_B and the network N'_{l_1} for the example problem introduced in Figure 9.1.

9.2 Constructing a Dominating Set

The decomposition of $\mathcal{F}$ into cells $\mathcal{C}(N_{l_1})$ will be used in this section to derive some useful properties of l_1-shortest permitted paths from arbitrary feasible points to the existing facilities. Since an extended network is used compared to that defined in Larson and Sadiq (1983), the following two results, which can also be found in this reference, can now be proven in a more straightforward way.

Corollary 9.1. *Let $Y \in (\mathcal{E}x \cup \mathcal{P}(\mathcal{B}))$ be an existing facility or an extreme point of a barrier, and let C be a cell in $\mathcal{C}(\mathcal{N}_{l_1})$. If Y is l_1-visible from some point in $\mathrm{int}(C)$, then Y is l_1-visible from all points in C.*

Proof. Since $\mathcal{N}_{l_1}$ is defined according to Definition 8.1, this result is an immediate consequence of Corollary 8.1 on page 125. □

Lemma 9.2. *Let $Ex_m \in \mathcal{E}x$ be an existing facility and let C be a cell in $\mathcal{C}(\mathcal{N}_{l_1})$ with $X \in C$. Then there exists an l_1-shortest permitted Ex_m-X path that passes through a corner point of C.*

Proof. Let $Ex_m \in \mathcal{E}x$ and let $X \in C$. Since the case that X is a corner point of C is trivial, assume that X lies in the interior of C or on a facet of C. Furthermore, let SP_m be an l_1-shortest permitted path connecting Ex_m and X and satisfying the barrier touching property of Lemma 2.3. Then there exists an intermediate point $I_m \in (\mathcal{E}x \cup \mathcal{P}(\mathcal{B}))$ on SP_m that is l_1-visible from X. Moreover, Corollary 8.2 on page 125 implies that SP_m and I_m can be chosen such that I_m is also l_1-visible from all points in C.

Let $x_1 := \min\{x : (x,y)^T \in C\}$, $x_2 := \max\{x : (x,y)^T \in C\}$, $y_1 := \min\{y : (x,y)^T \in C\}$, and $y_2 := \max\{y : (x,y)^T \in C\}$, and let $I_m = (a_m, b_m)^T$. Then a_m cannot be contained in the open interval (x_1, x_2), and b_m cannot be contained in the open interval (y_1, y_2), because otherwise a construction line would exist intersecting $\mathrm{int}(C)$.

Using additionally the fact that I_m is also l_1-visible from every corner point of C and that the corner points of C are l_1-visible from every point in C, an l_1-shortest permitted path connecting X and I_m that passes through a corner point of C can be constructed; see Figure 9.4 for an example. □

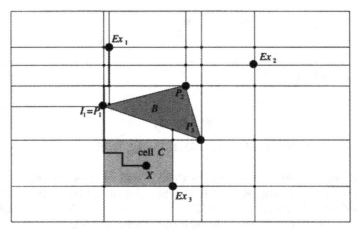

Figure 9.4. An example of an l_1-shortest permitted path connecting Ex_1 and X that passes through a corner point of the cell C.

Lemma 9.2 implies that we can always find l_1-shortest permitted paths with the following property: Every cell that is intersected by this path is entered through one corner point and left through a different corner point. However, there might exist cells $C \subseteq (\mathcal{F} \cap R_\mathcal{B})$ having two corner points C_1 and C_2 that are not connected by a network path of length $l_1(C_1, C_2)$ in N'_{l_1} even though they satisfy $l_{1,\mathcal{B}}(C_1, C_2) = l_1(C_1, C_2)$; see Figure 9.5 for an example.

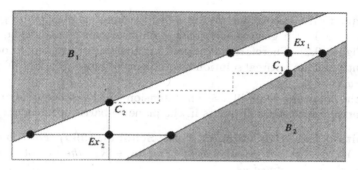

Figure 9.5. The two corner points C_1 and C_2 are not connected by a network path of length $l_1(C_1, C_2)$ in N'_{l_1}.

Extending N'_{l_1} by all those edges of length $l_1(C_1, C_2)$ that connect two corner points C_1 and C_2 of the same cell C that are not yet connected by a network path of length $l_1(C_1, C_2)$ leads to a network N''_{l_1} with the following property:

Corollary 9.2. *The length of an l_1-shortest permitted path between a corner point of a cell and an existing facility in $\mathcal{E}x$ is equal to the length of a shortest network path connecting the corresponding vertices in N''_{l_1}.*

Moreover, the length of an l_1-shortest permitted path between two existing facilities in $\mathcal{E}x$ is equal to the length of a shortest network path connecting the corresponding vertices in N''_{l_1}.

Observe that Corollary 9.2 can in general not be extended to points on lines or line segments of the grid $\mathcal{N}_{l_1}$ corresponding to points on edges of N''_{l_1}. This can be seen, for example, in Figure 9.5, where an l_1-shortest permitted path from points on the facet of B_2 bounding the cell C to the corner point C_2 is not represented in the corresponding network N''_{l_1}.

Thus in the case that an optimal solution of $1/P/(\mathcal{B} = N$ *convex polyhedra*$)/l_{1,\mathcal{B}}/$ max exists on $\mathcal{N}_{l_1}$, this solution cannot necessarily be found as an optimal solution of the corresponding network location problem $1/N''_{l_1}/ \bullet /d_{N''_{l_1}}(V, E)/$ max on N''_{l_1}. However, the network N''_{l_1} can be used in a way similar to that of the visibility graph in Section 4.2 (cf. Theorem 4.3 on page 52) to derive an improved upper bound for the optimal objective value of $1/P/(\mathcal{B} = N$ *convex polyhedra*$)/l_{1,\mathcal{B}}/$ max:

Corollary 9.3. *Let N''_{l_1} be the extension of N'_{l_1} as defined above. If $X^*_{N''_{l_1}}$ is an optimal solution of the network location problem $1/N''_{l_1}/\bullet/d_{N''_{l_1}}(V,E)/$ max on N''_{l_1}, then the point in the plane corresponding to the point $X^*_{N''_{l_1}}$ in the embedding of N''_{l_1} is feasible for $1/P/(\mathcal{B} = N$ convex polyhedra$)/l_{1,\mathcal{B}}/$ max and*

$$f_{\mathcal{B}}(X^*_{\mathcal{B}}) \le f_{\mathcal{B}}(X^*_{N''_{l_1}}) \le f_{N''_{l_1}}(X^*_{N''_{l_1}}).$$

Contrary to the case of the Weber objective function the bound given in Corollary 9.3 is in general not sharp (compare Larson and Sadiq (1983) and Figure 9.2). Therefore, additional arguments are needed in order to efficiently find an optimal solution to problems of type $1/P/(\mathcal{B} = N$ convex polyhedra$)/l_{1,\mathcal{B}}/$ max.

In the following we will use an extension of the concept of *weighted bisectors* between pairs of points in the plane to barrier problems:

Definition 9.1. (See Mitchell, 1992; Schöbel, 1999) *For two distinct points $Y_1, Y_2 \in \mathcal{F}$ with positive weights $w_1, w_2 \in \mathbb{R}_+$, the weighted bisector of Y_1 and Y_2 is defined as*

$$b(w_1 Y_1, w_2 Y_2) := \{X \in \mathcal{F} : w_1 l_{1,\mathcal{B}}(X, Y_1) = w_2 l_{1,\mathcal{B}}(X, Y_2)\}.$$

Moreover, if additionally two constants $d_1, d_2 \in \mathbb{R}_+$ are given, the weighted bisector of Y_1, d_1 and Y_2, d_2 is defined as

$$b(w_1(Y_1, d_1), w_2(Y_2, d_2))$$
$$:= \{X \in \mathcal{F} : w_1(l_{1,\mathcal{B}}(X, Y_1) + d_1) = w_2(l_{1,\mathcal{B}}(X, Y_2) + d_2)\}.$$

For an overview of the construction of bisectors in the unconstrained case we refer to Weißler (1999) and Icking et al. (1999).

A well-known result for center problems that also applies to center problems with barriers is that every optimal solution has to be located on the weighted bisector of two existing facilities. Otherwise, the objective value could be improved by moving the new location towards the existing facility at maximum weighted distance. Note that therefore an optimal solution can always be found as a point on the farthest-point Voronoi diagram with respect to the existing facilities, taking into account the barrier regions (see Okabe et al. (1992) and Shamos and Hoey (1975) for the unconstrained case).

Since the construction of the farthest-point Voronoi diagram is difficult in the presence of barriers, a solution strategy to solve $1/P/(\mathcal{B} = N$ convex polyhedra$)/l_{1,\mathcal{B}}/$ max will be developed that avoids the construction of all the weighted bisectors of the existing facilities or the corresponding Voronoi diagram.

In the following we will distinguish two different scenarios:

Scenario 1: There exists an optimal solution $X_{\mathcal{B}}^* \in \mathcal{X}_{\mathcal{B}}^*$ of $1/P/(\mathcal{B} = N$ $convex$ $polyhedra)/l_{1,\mathcal{B}}/$ max with objective value $z_{\mathcal{B}}^*$ such that $X_{\mathcal{B}}^* \in N_{l_1}''$ and

$$w_p l_{1,\mathcal{B}}(X_{\mathcal{B}}^*, Ex_p) = z_{\mathcal{B}}^* = w_q l_{1,\mathcal{B}}(X_{\mathcal{B}}^*, Ex_q)$$

is satisfied for *exactly two* different existing facilities $Ex_p, Ex_q \in \mathcal{E}x$. In this case $X_{\mathcal{B}}^*$ must lie on the intersection of the weighted bisector $b(w_p Ex_p, w_q Ex_q)$ of Ex_p and Ex_q with the network N_{l_1}''.

Scenario 2: Otherwise, there does not exist an optimal solution $X_{\mathcal{B}}^*$ of $1/P/(\mathcal{B} = N$ $convex$ $polyhedra)/l_{1,\mathcal{B}}/$ max with optimal objective value $z_{\mathcal{B}}^*$ on N_{l_1}'' that is of maximal weighted distance $z_{\mathcal{B}}^*$ from only two existing facilities in $\mathcal{E}x$. Then an optimal solution $X_{\mathcal{B}}^*$ may exist in the interior of a cell $C \subseteq (\mathcal{F} \cap R_{\mathcal{B}})$ that lies on only one weighted bisector $b(w_i Ex_i, w_j Ex_j)$ of two different existing facilities $Ex_i, Ex_j \in \mathcal{E}x$. An example for this situation is given in Figure 9.6. However, the following theorem proves that in this case an optimal solution of $1/P/(\mathcal{B} = N$ $convex$ $polyhedra)/$ $l_{1,\mathcal{B}}/$ max can be found in the intersection of two weighted bisectors $b(w_i Ex_i, w_j Ex_j)$ and $b(w_j Ex_j, w_k Ex_k)$, determined by three distinct existing facilities $Ex_i, Ex_j, Ex_k \in \mathcal{E}x$.

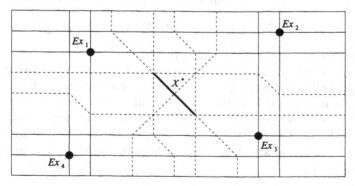

Figure 9.6. An example problem with four existing facilities having equal weights $w_m = 1$, $m = 1, \ldots, 4$. The weighted bisectors of all pairs of existing facilities are represented by dashed lines. Observe that the set of optimal solutions of $1/P/ \bullet /l_1/$ max contains solutions in the interior of a cell that lie only on the weighted bisector $b(Ex_2, Ex_4)$. However, there does not exist an optimal solution on the network N_{l_1}''.

Theorem 9.2. *Let $\mathcal{X}_{\mathcal{B}}^*$ be the set of optimal solutions of $1/P/(\mathcal{B} = N$ convex polyhedra$)/l_{1,\mathcal{B}}/$ max according to Scenario 2; i.e., there does not exist $X_{\mathcal{B}}^* \in \mathcal{X}_{\mathcal{B}}^*$ with $X_{\mathcal{B}}^* \in N_{l_1}''$ such that $X_{\mathcal{B}}^*$ is at maximum weighted distance from only two of the existing facilities in $\mathcal{E}x$.*

Let $z_{\mathcal{B}}^*$ be the optimal objective value of the problem. Then there exists at least one optimal solution $X_{\mathcal{B}}^* \in \mathcal{X}_{\mathcal{B}}^*$ that has the weighted distance $z_{\mathcal{B}}^*$ from at least three different existing facilities in $\mathcal{E}x$.

Proof. First suppose that there exists $X_{\mathcal{B}}^* \in \mathcal{X}_{\mathcal{B}}^*$ with $X_{\mathcal{B}}^* \in N_{l_1}''$. Then the fact that every optimal solution of $1/P/(\mathcal{B} = N$ convex polyhedra$)/l_{1,\mathcal{B}}/$ max has to lie on the weighted bisector of at least two existing facilities and the assumption that there exists no optimal solution on N_{l_1}'' being at maximum weighted distance from exactly two existing facilities implies the result.

Now suppose that $X_{\mathcal{B}}^*$ is an optimal solution of $1/P/(\mathcal{B} = N$ convex polyhedra$)/l_{1,\mathcal{B}}/$ max and that C is a cell such that $X_{\mathcal{B}}^* \in \text{int}(C)$. Then Theorem 9.1 implies that $C \subseteq R_{\mathcal{B}} \cap \mathcal{F}$. Obviously, there exist at least two existing facilities Ex_i and Ex_j in $\mathcal{E}x$ with $z_{\mathcal{B}}^* = w_i l_{1,\mathcal{B}}(X_{\mathcal{B}}^*, Ex_i) = w_j l_{1,\mathcal{B}}(X_{\mathcal{B}}^*, Ex_j)$. Furthermore, let C_i (and C_j, respectively) be a corner point of C such that there exists an l_1-shortest permitted path connecting Ex_i and $X_{\mathcal{B}}^*$ (Ex_j and $X_{\mathcal{B}}^*$, respectively) passing through C_i (C_j, respectively); see Lemma 9.2.

Now assume that there is no existing facility $Ex_m \in \mathcal{E}x$ other than Ex_i and Ex_j such that $w_m l_{1,\mathcal{B}}(X_{\mathcal{B}}^*, Ex_m) = z_{\mathcal{B}}^*$. Then $C_i \neq C_j$, since otherwise the objective value $z_{\mathcal{B}}^*$ could be improved by moving X^* towards C_i in C.

Defining $d_i := l_{1,\mathcal{B}}(C_i, Ex_i)$ and $d_j := l_{1,\mathcal{B}}(C_j, Ex_j)$ we obtain that

$$w_i(l_1(X_{\mathcal{B}}^*, C_i) + d_i) = w_j(l_1(X_{\mathcal{B}}^*, C_j) + d_j) = z_{\mathcal{B}}^*$$

and thus $X_{\mathcal{B}}^* \in b(w_i(C_i, d_i), w_j(C_j, d_j)) \cap C$.

Due to the optimality of $X_{\mathcal{B}}^*$ and to the fact that the weighted distance from $X_{\mathcal{B}}^*$ to all other existing facilities is strictly less than $z_{\mathcal{B}}^*$, $X_{\mathcal{B}}^*$ has to be located on an l_1-shortest permitted path connecting C_i and C_j in C. Since there also exists an l_1-shortest permitted path connecting C_i and C_j on the network N_{l_1}'', there exists a point $X_{N_{l_1}''} \in N_{l_1}''$ (not necessarily a node, i.e., $X_{N_{l_1}''}$ may lie on an edge) different from $X_{\mathcal{B}}^*$ on this path (and in the cell C) such that

$$w_i\left(l_1(X_{N_{l_1}''}, C_i) + d_i\right) = w_j\left(l_1(X_{N_{l_1}''}, C_j) + d_j\right) = z_{\mathcal{B}}^*.$$

Thus $X_{N_{l_1}''} \neq X_{\mathcal{B}}^*$ is also a point on the weighted bisector of C_i, d_i and C_j, d_j; i.e., $X_{N_{l_1}''} \in (b(w_i(C_i, d_i), w_j(C_j, d_j)) \cap C)$. Since C is convex, all points on the line segment

$$\overline{X_{\mathcal{B}}^*, X_{N_{l_1}''}} := \left\{ X \in C : X = \lambda X_{\mathcal{B}}^* + (1 - \lambda) X_{N_{l_1}''}, \ \lambda \in [0, 1] \right\}$$

connecting $X_{\mathcal{B}}^*$ and $X_{N_{l_1}''}$ lie in C. Furthermore, for $m \in \{i, j\}$ all points $X \in \overline{X_{\mathcal{B}}^*, X_{N_{l_1}''}}$ satisfy

$$
\begin{aligned}
w_m l_{1,\mathcal{B}}(X, C_m) &= w_m l_1(X, C_m) \\
&= w_m l_1(\lambda X_{\mathcal{B}}^* + (1-\lambda) X_{N_{l_1}''}, C_m) \\
&= w_m \left(\lambda l_1(X_{\mathcal{B}}^*, C_m) + (1-\lambda) l_1(X_{N_{l_1}''}, C_m) \right) \\
&= z_{\mathcal{B}}^* - w_m d_m.
\end{aligned}
$$

Thus $X_{\mathcal{B}}^* \in \overline{X_{\mathcal{B}}^*, X_{N_{l_1}''}}$ can be moved along the line segment $\overline{X_{\mathcal{B}}^*, X_{N_{l_1}''}}$ (which is part of the bisector $b(w_i(C_i, d_i), w_j(C_j, d_j)) \cap C$) in the cell C without increasing the weighted distance to Ex_i and Ex_j until the weighted distance to some other existing facility $Ex_m \in \mathcal{E}x$ equals $z_{\mathcal{B}}^*$, $m \notin \{i, j\}$, or until it reaches the boundary of C. In the latter case, the point $X_{N_{l_1}''}$ is an optimal solution according to Scenario 1, a case that is excluded by the assumption (see Figure 9.7 for an example). □

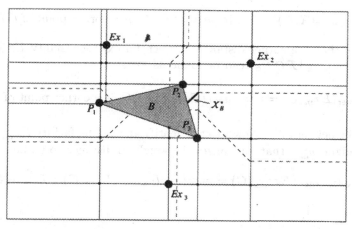

Figure 9.7. The weighted bisectors $b(Ex_i, Ex_j)$ for the example problem introduced in Figure 9.1, where the weights are chosen equal to 1. Note that there exist optimal solutions on N_{l_1}'', but not at an intersection of weighted bisectors.

Let $DS\left(N_{l_1}''\right)$ denote the set of those points in N_{l_1}'' that are located on the intersection of a weighted bisector $b(w_p Ex_p, w_q Ex_q)$ of two existing facilities with the network N_{l_1}''. Obviously, $DS\left(N_{l_1}''\right)$ contains an optimal solution of $1/P/(\mathcal{B} = N\,convex\,polyhedra)/l_{1,\mathcal{B}}/\max$ in case of Scenario 1.

Similarly, let $DS\left(\mathcal{C}\left(\mathcal{N}_{l_1}\right)\right)$ denote the set of points in cells $C \in \mathcal{C}\left(\mathcal{N}_{l_1}\right)$ that are located on the intersection of at least two weighted bisectors $b(w_i Ex_i, w_j Ex_j)$ and $b(w_j Ex_j, w_k Ex_k)$ determined by three distinct existing facilities $Ex_i, Ex_j, Ex_k \in \mathcal{E}x$. Observe that

$$
\begin{aligned}
&b(w_i Ex_i, w_j Ex_j) \cap b(w_j Ex_j, w_k Ex_k) \\
&= \ b(w_i Ex_i, w_j Ex_j) \cap b(w_i Ex_i, w_k Ex_k) \\
&= \ b(w_i Ex_i, w_k Ex_k) \cap b(w_j Ex_j, w_k Ex_k).
\end{aligned}
$$

Consequently, Theorem 9.2 enables us to construct a dominating set $DS \subseteq \mathcal{F}$, containing at least one optimal solution of $1/P/(\mathcal{B} = N \text{ convex polyhedra})/l_{1,\mathcal{B}}/\max$. This dominating set can be defined as the union of the two sets $DS\left(N''_{l_1}\right)$ and $DS\left(\mathcal{C}\left(\mathcal{N}_{l_1}\right)\right)$; i.e.,

$$
DS := DS\left(N''_{l_1}\right) \cup DS\left(\mathcal{C}\left(\mathcal{N}_{l_1}\right)\right).
$$

Before we develop an algorithm to solve $1/P/(\mathcal{B} = N \text{ convex polyhedra})/l_{1,\mathcal{B}}/\max$ based on the dominating set DS, we will first reduce the set DS by further exploiting the particular structure of the problem.

Consider an arbitrary cell $C \in \mathcal{C}\left(\mathcal{N}_{l_1}\right)$ and let the *corner distance* between a point $X \in \mathcal{F} \setminus C$ and the cell C be defined as

$$
l_{1,\text{corner}}(X, C) := \min\{l_{1,\mathcal{B}}(X, C_i) \ : \ C_i \text{ is a corner point of } C\}.
$$

Observe that the corner distance between an existing facility $Ex_m \in \mathcal{E}x$ and a cell $C \subseteq (\mathcal{F} \cap R_{\mathcal{B}})$ can be found as

$$
l_{1,\text{corner}}(Ex_m, C) = \min\left\{ d_{N''_{l_1}}(Ex_m, C_i) \ : \ C_i \text{ is a corner point of } C \right\},
$$

using the network N''_{l_1}. Now we can identify an existing facility $Ex^C_{\max} \in \mathcal{E}x$ with weight $w^C_{\max}$ that maximizes the weighted distance to C; i.e.,

$$
w^C_{\max} \, l_{1,\text{corner}}\left(Ex^C_{\max}, C\right) = \max\{w_m \, l_{1,\text{corner}}(Ex_m, C) \ : \ Ex_m \in \mathcal{E}x\}.
$$

Furthermore, let

$$
|C| := \max\{l_1(C_i, C_j) \ : \ C_i \text{ and } C_j \text{ are corner points of } C\}
$$

denote the maximal distance between two corner points of a cell $C \in \mathcal{C}\left(\mathcal{N}_{l_1}\right)$.

Lemma 9.3. *Let $\mathcal{X}^*_{\mathcal{B}}$ be the set of optimal solutions of $1/P/(\mathcal{B} = N \text{ convex polyhedra})/l_{1,\mathcal{B}}/\max$ and let $z^*_{\mathcal{B}}$ be the optimal objective value of the problem. Then every optimal solution $X^*_{\mathcal{B}} \in \mathcal{X}^*_{\mathcal{B}}$ in a cell $C \in \mathcal{C}\left(\mathcal{N}_{l_1}\right)$ lies on the weighted bisector of at least two different existing facilities $Ex_i, Ex_j \in \mathcal{E}x$ satisfying*

$$
w_p \, l_{1,\text{corner}}(Ex_p, C) + w_p \, |C| \geq w^C_{\max} \, l_{1,\text{corner}}\left(Ex^C_{\max}, C\right), \qquad p = i, j.
$$

Proof. Recall that every optimal solution $X_{\mathcal{B}}^* \in \mathcal{X}_{\mathcal{B}}^*$ of $1/P/(\mathcal{B} = N \; convex$ $polyhedra)/l_{1,\mathcal{B}}/$ max lies on the weighted bisector of at least two existing facilities $Ex_i, Ex_j \in \mathcal{E}x$ with $z_{\mathcal{B}}^* = w_p l_{1,\mathcal{B}}(X_{\mathcal{B}}^*, Ex_p)$ for $p = i, j$.

Let $C \in \mathcal{C}(\mathcal{N}_{l_1})$ be a cell with $X_{\mathcal{B}}^* \in C$. Then

$$z_{\mathcal{B}}^* \geq w_{\max}^C \, l_{1,\text{corner}}\left(Ex_{\max}^C, C\right).$$

On the other hand,

$$z_{\mathcal{B}}^* = w_p \, l_{1,\mathcal{B}}(X_{\mathcal{B}}^*, Ex_p) \leq w_p \, l_{1,\text{corner}}(Ex_p, C) + w_p \, |C|, \qquad p = i, j,$$

which implies the result. $\square$

Thus with respect to each cell it is sufficient to consider only those existing facilities that satisfy the distance requirement given in Lemma 9.3. Especially in applications with a large number of uniformly distributed existing facilities, this result leads to a significant reduction of intersections of weighted bisectors that have to be considered.

Summarizing the results above, a dominating set result for an optimal solution of $1/P/(\mathcal{B} = N \; convex \; polyhedra)/l_{1,\mathcal{B}}/$ max can be stated in the following way:

Theorem 9.3. *Let DS be a set of points in $\mathcal{F}$ consisting, in all cells $C \subseteq R_{\mathcal{B}} \cap \mathcal{F}$, of*

(i) *the intersection points of the network N_{l_1}'' with the weighted bisector determined by at least two different existing facilities $Ex_i, Ex_j \in \mathcal{E}x$, and*

(ii) *the intersection points of at least two weighted bisectors determined by three distinct existing facilities Ex_i, Ex_j, Ex_k,*

where only those existing facilities $Ex_p \in \mathcal{E}x$ are considered in (i) and (ii) that satisfy

$$w_p \, l_{1,\text{corner}}(Ex_p, C) + w_p \, |C| \geq w_{\max}^C \, l_{1,\text{corner}}\left(Ex_{\max}^C, C\right).$$

Then DS contains at least one optimal solution of $1/P/(\mathcal{B} = N \; convex \; polyhedra)/l_{1,\mathcal{B}}/$ max.

Note that the dominating set *DS* of Theorem 9.3 is in general not finite, since the set of intersection points of two weighted bisectors with respect to the l_1 metric is not necessarily a finite set. An example for this situation is given in Figure 9.8.

9.3   Algorithmic Consequences

As was shown in the previous sections, in the case of the center objective function it is not sufficient to consider intersection points of the grid $\mathcal{N}_{l_1}$,

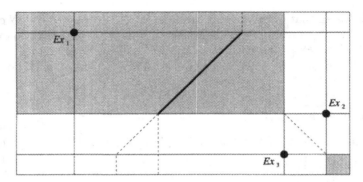

Figure 9.8. The weighted bisectors $b(Ex_1. Ex_2)$. $b(Ex_2. Ex_3)$, and $b(Ex_1, Ex_3)$ intersect in the line segment represented by a bold line. Note that the shaded regions are part of the weighted bisector $b(Ex_2. Ex_3)$.

as was the case for the Weber objective function. Moreover. the problem is not equivalent to a network location problem on N_{l_1}'' even though this was proven for the corresponding Weber problem. Therefore, it is necessary to consider additional points in the intersections of specific weighted bisectors between existing facilities in order to find an optimal solution of the center problem with barriers. An example problem where the unique optimal solution lies in the intersection of weighted bisectors and not on the grid N_{l_1}'' is given in Figure 9.9. Observe also that. for example. the weighted bisector $b(Ex_1, Ex_3)$ has a breakpoint in the interior of a cell in this example.

The following algorithm for the solution of center problems with barriers and the Manhattan metric is based on the construction of weighted bisectors between pairs of existing facilities. yielding a dominating set DS for an optimal solution of $1/P/(\mathcal{B} = N\ convex\ polyhedra)/l_{1.\mathcal{B}}/\max$ as discussed above. Observe that the determination of the individual candidate sets $DS\left(N_{l_1}''\right)$ and $DS\left(C\left(N_{l_1}\right)\right)$ can be combined, since both candidate sets use the same weighted bisectors between pairs of existing facilities.

Algorithm 9.1. (Bisector Algorithm)

Input: *Location problem* $1/P/(\mathcal{B} = N\ convex\ polyhedra)/l_{1.\mathcal{B}}/\max$.

Step 1:   *Construct* $\mathcal{N}_{l_1}$, $R_{\mathcal{B}}$. *and* N_{l_1}''.

Step 2:   *Find a dominating set DS by determining the weighted bisectors* $b(w_i Ex_i, w_j Ex_j) \cap R_{\mathcal{B}}$ *between all pairs of existing facilities, and the intersection of*

   (a) *all pairs of weighted bisectors. determined by three existing facilities at a time*

   (b) *and of weighted bisectors with the network* N_{l_1}''.

Step 3:   *From the candidate solutions determined in Steps 2(a) and (b),*
            find $z_{\mathcal{B}}^* := \min\{f_{\mathcal{B}}(X) : X \in DS\}$ *and*
            $\mathcal{X}_{\mathcal{B}} := \arg\min\{f_{\mathcal{B}}(X) : X \in DS\}.$

Output: *A subset $\mathcal{X}_{\mathcal{B}}$ of the set of optimal solutions $\mathcal{X}_{\mathcal{B}}^*$ of $1/P/(\mathcal{B} = N$ convex polyhedra$)/l_{1,\mathcal{B}}/\max$ and the optimal objective value $z_{\mathcal{B}}^*$.*

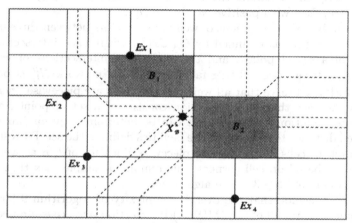

Figure 9.9. An example problem with four existing facilities having equal weights $w_m = 1$, $m = 1, \ldots, 4$. The weighted bisectors of all pairs of existing facilities are represented by dashed lines.

In Step 1 of Algorithm 9.1 the fundamental data structures are implemented, and a network N_{l_1}'' of the same asymptotic size as the network used in Larson and Sadiq (1983) for the corresponding Weber problem is constructed. In particular, the number of vertices, edges, and cells of the original network N_{l_1} can be bounded by $O\left((|\mathcal{E}x| + |\mathcal{P}(\mathcal{B})|)^2\right)$ using simple planarity arguments (cf. page 61). Since $N_{l_1}' \subseteq N_{l_1}$ and since the number of edges added in the transition from N_{l_1}' to N_{l_1}'' can be bounded by a constant within each cell, the same bounds also hold for the extended network N_{l_1}''.

In contrast to the case of the median objective function, the additional determination of weighted bisectors is needed in the case of the center objective function in Step 2 of the algorithm. These bisectors can be found by adapting an algorithm of Mitchell (1992) that is based on the determination of l_1-shortest permitted paths in the presence of polyhedral barriers. The presented algorithm extends a "continuous Dijkstra" technique of propagating a "wavefront" from a given source node s towards a termination node t. The propagation is implemented based on "dragged segments," that is, line segments that make up portions of the wavefront and that are dragged in the southeast, northeast, northwest, and southwest directions, respectively, according to the unit ball of the Manhattan metric. Mitchell (1992) showed that this approach can also be used for the determination of bisectors between pairs of points. In particular, the bisector $b(Ex_i, Ex_j)$ between two existing facilities can be constructed in time $O(|\mathcal{P}(\mathcal{B})| \cdot \log |\mathcal{P}(\mathcal{B})|)$ by initiating one wavefront at each of the two facilities. If the weight of the respective facility is associated with its wavefront (and with each of the corresponding dragged segments whose data structure includes the information about the segment's distance from its source node), this algorithm can be adapted to handle the more general case of a weighted bisector $b(w_i Ex_i, w_j Ex_j)$ with positive weights w_i, w_j.

The number of intersections of weighted bisectors between three existing facilities at a time is bounded by $O(|\mathcal{E}x|^3)$, where each intersection may consist of a set of points, and the number of intersections of weighted bisectors between two existing facilities and the network N''_{l_1} is bounded by $O(|\mathcal{E}x|^2)$. Observe that all weighted bisectors are piecewise linear, and that in the case that an intersection consists of a set of points, a "best candidate" solution within this set can be determined by minimizing the weighted distance to only one of the facilities defining the set. By restricting this search to cells and using the piecewise linearity of $l_{1,\mathcal{B}}$ as well as its description based on cell corners, this remains a simple task that can be solved as part of Step 3 of the algorithm.

We can conclude that the overall complexity of Algorithm 9.1 remains polynomial even if all of the $O(|\mathcal{E}x|^3)$ intersection sets are examined. Note that the reductions of the dominating set due to Theorem 9.1 and Lemma 9.3 are not yet reflected in this discussion.

It is an interesting open question whether the wavefront approach used in Mitchell (1992) could be further combined with the ideas discussed in the previous sections. An intuitive idea for an efficient solution concept for center problems with barriers would be to start an "elementary wave" at each existing facility and propagate the resulting wavefronts around the barrier sets until the first point of common intersection is found. This "center point" can be identified as that point in $\mathcal{F}$ where all the waves emanating from the existing facilities first intersect.

A first promising result in this direction was obtained in the two recent master's projects Frieß (1999) and Sprau (1999). In these projects, prototype algorithms were implemented that allow the solution of simple example problems with the Manhattan metric and with the Euclidean metric where all existing facilities have equal weights. This interesting approach should be further developed in future research.

9.4   Extension to Block Norms

In this section possible extensions of the results developed in the previous sections to other, more general, distance functions are discussed.

Let γ be a block norm as defined in Definition 1.6 on page 7 with exactly four fundamental vectors (see Figure 1.7 on page 12 for an example). To distinguish γ from the Manhattan metric with the fundamental vectors $v^1, \ldots, v^4$, we will denote the fundamental vectors of γ by $u^1, \ldots, u^4$ (in clockwise order), where

$$u^1 = \left(u_1^1, u_2^1\right)^T, \quad u^2 = \left(u_1^2, u_2^2\right)^T, \quad u^3 = \left(u_1^3, u_2^3\right)^T, \quad u^4 = \left(u_1^4, u_2^4\right)^T$$

such that $u^3 = -u^1$ and $u^4 = -u^2$.

Under these assumptions, a linear transformation T satisfying $T\left(u^1\right) = Tu^1 = v^1$ and $T\left(u^2\right) = Tu^2 = v^2$ can be defined according to (1.26) on page 13. Using Lemma 2.7 on page 37, the following generalizations of the concepts developed in the previous section for the Manhattan metric can be proven:

Theorem 9.4. *Let γ be a block norm with four fundamental vectors and let T be a linear transformation defined according to (1.26). Furthermore, let $\mathcal{B}$ be the union of a finite set of compact, convex, pairwise disjoint and polyhedral barriers, and let $\mathcal{E}x = \{Ex_m : m \in \mathcal{M}\}$ be a set of existing facilities in $\mathcal{F} = \mathbb{R}^2 \setminus \mathrm{int}(\mathcal{B})$.*

Then $X_{\mathcal{B}}^*$ is an optimal solution of $1/P/(\mathcal{B} = N \text{ convex polyhedra})/\gamma_\mathcal{B}/$ max with existing facilities $\mathcal{E}x$ if and only if $T(X_{\mathcal{B}}^*)$ is an optimal solution of the transformed problem $1/P/T(\mathcal{B})/l_{1,T(\mathcal{B})}/$ max with existing facilities $T(\mathcal{E}x) := \{T(Ex_m) : m \in \mathcal{M}\}$ and barriers $T(\mathcal{B})$.

Proof. This result follows directly from Lemma 2.7   □

Even though the proof of the existence of a dominating set was somewhat technical for the case of the Manhattan metric, this section easily showed how valuable these results are in the more general case of block norms.

Observe also that a generalization of the above results to block norms with more than four fundamental vectors is not possible, since Lemma 9.2 is in general not true in this case. An example can be seen in Figure 8.4 on page 127, where no γ-shortest permitted path from X to Ex_3 exists that passes through a corner point of the grid $\mathcal{N}_\gamma$.

10

Multicriteria Location Problems with Polyhedral Barriers

Multiple objective location problems involve locating a new facility with respect to a given set of existing facilities so that a vector of performance criteria is optimized. In a major part of location planning, especially in regional and social planning, several decision-makers with different priorities are involved in the locational decision, which causes a growing need for efficient solution strategies for location models, including multiple objective functions; see, for example, Kuhpfahl et al. (1995).

Theoretical results and algorithmic approaches for multicriteria location problems without barriers have been discussed by Chalmet et al. (1981), Durier (1990), Ehrgott et al. (1999), Gerth and Pöhler (1988), Hamacher and Nickel (1993, 1996), Nickel (1994, 1995, 1997), McGinnis and White (1978), Tammer and Tammer (1991), Wanka (1991, 1995), Wendell and Hurter (1973), and Wendell et al. (1973), among others. For an overview and a classification of different types of multicriteria optimization problems we refer to Ehrgott (2000).

The primary goal in this chapter is the analytical determination of the efficient set of multiple objective location problems with polyhedral barriers. The structure of the efficient set is first examined in order to motivate the design of special algorithms. The theoretical analysis will show that the original nonconvex problem can be decomposed into a series of multiple objective convex subproblems, similar to the decomposition presented in Chapter 5. A less general formulation of the results given in this chapter can also be found in Klamroth and Wiecek (2002).

The problem we consider is based on a general multiple objective location model in the plane $\mathbb{R}^2$ that can be viewed as a generalization of (3.1) to the multicriteria case. It can be formulated as

$$\text{vmin} \quad (f_1(X), \ldots, f_Q(X))^T \qquad (Q \geq 2)$$
$$\text{s.t.} \quad X \in \mathbb{R}^2 \tag{10.1}$$

and has the classification $1/P/ \bullet /d/Q - f \ convex$.

The Q individual criteria measure the performance of a locational decision in the plane $\mathbb{R}^2$ with respect to a finite set of existing facilities $\mathcal{E}x = \{Ex_1, Ex_2, \ldots, Ex_M\}$. Each objective is given as a convex function depending on the distances between the new facility and the existing facilities in $\mathcal{E}x$. Thus

$$f_q(X) = f_q(d_q(X, Ex_1), \ldots, d_q(X, Ex_M)), \qquad q = 1, \ldots, Q, \tag{10.2}$$

where we additionally assume that f_q is a convex and nondecreasing function of the distances $d_q(X, Ex_1), \ldots, d_q(X, Ex_M)$, $q = 1, \ldots, Q$.

Since each decision-maker may consider different ways of transportation, distances may be measured differently in each objective function. Thus for each criterion $q \in \{1, \ldots, Q\}$, d_q is an arbitrary metric induced by a norm.

Solving (10.1) is understood as generating its efficient (Pareto) solutions. A feasible solution $X_E \in \mathbb{R}^2$ is said to be an *efficient solution* of (10.1) if there is no other solution $X \in \mathbb{R}^2$ such that $f(X) < f(X_E)$; i.e.,

$$\forall q \in \{1, \ldots, Q\} \qquad f_q(X) \leq f_q(X_E),$$
$$\exists q \in \{1, \ldots, Q\} \quad \text{s.t.} \quad f_q(X) < f_q(X_E). \tag{10.3}$$

Let $\mathcal{X}_E$ denote the set of efficient solutions of (10.1) and let $\mathcal{Y}_E$ denote the image of $\mathcal{X}_E$ in the objective space; that is, $\mathcal{Y}_E = f(\mathcal{X}_E)$, where $f = (f_1, \ldots, f_Q)^T$. The set $\mathcal{Y}_E$ is referred to as the set of *nondominated points* of (10.1).

When each objective function of (10.1) is minimized individually over $\mathbb{R}^2$, the set of optimal solutions, denoted by $\mathcal{X}_q$, is obtained:

$$\mathcal{X}_q = \arg \min_{X \in \mathbb{R}^2} \{f_q(X)\}, \qquad q = 1, \ldots, Q.$$

We also define the *ideal point* $U = (u_1, \ldots, u_Q)^T$, where $u_q = \min_{X \in \mathbb{R}^2} f_q(X)$; i.e., $u_q = f_q(\mathcal{X}_q)$, $q = 1, \ldots, Q$.

The multiple objective location problem with barriers is a special case of (10.1), where the travel distances d_q (compare (10.2)) are lengthened due to one or several barriers in the plane.

For the union of a given finite set of convex, closed, polyhedral, and pairwise disjoint barrier sets $\mathcal{B} = \bigcup_{i=1}^N B_i$ in $\mathbb{R}^2$ the feasible region $\mathcal{F} = \mathbb{R}^2 \setminus \text{int}(\mathcal{B})$ is that region where new facilities can be located. Furthermore, we denote the length of a d_q-shortest permitted X-Y path according to Definition 2.2 by $d_{q,\mathcal{B}}(X, Y)$, $q = 1, \ldots, Q$.

Thus (10.1) can be restated as a *multiple objective location problem with convex, polyhedral barriers* $1/P/(\mathcal{B} = N\ convex\ polyhedra)/d_\mathcal{B}/Q - f\ convex$:

$$\text{vmin} \quad (f_{1,\mathcal{B}}(X), \ldots, f_{Q,\mathcal{B}}(X))^T$$
$$\text{s.t.} \quad X \in \mathcal{F} \tag{10.4}$$

with the individual objective functions given by

$$f_{q,\mathcal{B}}(X) = f(d_{q,\mathcal{B}}(X, Ex_1), \ldots, d_{q,\mathcal{B}}(X, Ex_M)), \qquad q = 1, \ldots, Q. \tag{10.5}$$

The set of efficient solutions of (10.4) is denoted by $\mathcal{X}_{E,\mathcal{B}}$, and the set of nondominated points of (10.4) is denoted by $\mathcal{Y}_{E,\mathcal{B}}$.

Observe that the individual objective functions of (10.4) are in general nonconvex. Consequently, the multiple objective problem may not have features possessed by convex multiple objective problems. In general, the efficient set $\mathcal{X}_{E,\mathcal{B}}$ may not be connected. Moreover, the set $\mathcal{Y}_{E,\mathcal{B}} + \mathbb{R}_{\geq}^2$ may be nonconvex (that is, one may encounter nondominated points in a duality gap), and the set $\mathcal{Y}_{E,\mathcal{B}}$ may be nonconnected; see Bitran and Magnanti (1979). For a detailed discussion of conditions for the connectedness of the efficient set of a multiple objective problem we refer to Luc (1989).

Since (10.4) is a nonconvex multiple objective problem, it may feature globally as well as locally efficient solutions that can be found by means of some suitable scalarizations specially developed to handle nonconvexity. All the globally efficient solutions can be found by means of the lexicographic weighted Chebyshev approach (see Steuer, 1986), while the locally efficient solutions can be generated using the augmented Lagrangian approach (see TenHuisen and Wiecek, 1996). Other scalarization approaches that can be applied are discussed, for example, in Kaliszewski (1994). In order to avoid treating (10.4) in this general methodological framework and to obtain specific and more effective approaches, we focus on the special case of convex polyhedral barriers.

In Section 10.1 we show that $1/P/(\mathcal{B} = N\ convex\ polyhedra)/d_\mathcal{B}/Q - f\ convex$ has a special structure that allows the development of conceptual results and specific approaches to finding the efficient solutions. In Section 10.2 an algorithm is proposed for the biobjective case, i.e., for the problem $1/P/(\mathcal{B} = N\ convex\ polyhedra)/d_\mathcal{B}/2 - f\ convex$, and the components of the algorithm are discussed for different measures of distance. Section 10.3 includes an illustrative example.

10.1   Properties of the Objective Function

Since we assumed that in each of the Q criteria a metric d_q induced by a norm is used to measure distances, Corollary 2.3 on page 34 is also applicable to the distance functions $d_{q,\mathcal{B}}$, $q = 1, \ldots, Q$. Thus the concept of intermediate points that was used in the single-objective case in Chapter 5 can be generalized to the multiple objective case. However, since we allow different distance measures in the Q objective functions, the visibility grid $\mathcal{G}_d$ (cf. Definition 5.2 on page 60) has to be replaced by a *generalized visibility grid* defined by all d_q-shadows of the existing facilities and the barriers, $q = 1, \ldots, Q$:

$$\mathcal{G}_{d_1,\ldots,d_Q} := \left(\bigcup_{q=1,\ldots,Q} \bigcup_{X \in \mathcal{E}x \cup \mathcal{P}(\mathcal{B})} \partial(\text{shadow}_{d_q}(X)) \right) \cup \mathcal{F}(\mathcal{B}). \qquad (10.6)$$

The set of cells of $\mathcal{G}_{d_1,\ldots,d_Q}$, i.e., the set of all polyhedra induced by $\mathcal{G}_{d_1,\ldots,d_Q}$ the nonempty interior of which is not intersected by a line segment in $\mathcal{G}_{d_1,\ldots,d_Q}$, is denoted by $\mathcal{C}(\mathcal{G}_{d_1,\ldots,d_Q})$. Note that using this definition, Lemmas 5.1 and 5.2 transfer to the grid $\mathcal{G}_{d_1,\ldots,d_Q}$.

Alternatively, based on Lemmas 2.2 and 2.3, the grid $\mathcal{G}_{l_2}$ can be used for all distances $d_1, \ldots, d_Q$. Since the latter approach generally yields a grid with a smaller number of cells, particularly in the case of many different distance measures, we will focus on this approach in the following. Nevertheless, the results of this section can be transferred to the case that the grid $\mathcal{G}_{d_1,\ldots,d_Q}$ is used to decompose the feasible region into a finite set of cells.

As with to Lemma 5.3 on page 64, the objective function of (10.4) can be reformulated with respect to each criterion $q \in \{1, \ldots, Q\}$:

Lemma 10.1. *Let $d = (d_1, \ldots, d_Q)^T$ be a vector of metrics induced by norms, let $C \in \mathcal{C}(\mathcal{G}_{l_2})$ be a cell, and let $X \in C$. Then for each existing facility $Ex_m \in \mathcal{E}x$ there exist intermediate points $I_{q,m} \in \mathcal{E}x \cup \mathcal{P}(\mathcal{B})$, $q = 1, \ldots, Q$, such that*

$$\begin{pmatrix} f_{1,\mathcal{B}}(X) \\ \vdots \\ f_{Q,\mathcal{B}}(X) \end{pmatrix} = \begin{pmatrix} F_{1,X}(X) \\ \vdots \\ F_{Q,X}(X) \end{pmatrix}. \qquad (10.7)$$

where for $q = 1, \ldots, Q$,

$$F_{q,X}(Y) := f_q(d_q(Y, I_{q,1}) + d_{q,\mathcal{B}}(I_{q,1}, Ex_1), \\ \ldots, d_q(Y, I_{q,M}) + d_{q,\mathcal{B}}(I_{q,M}, Ex_M)). \qquad (10.8)$$

Here $I_{q,m}$ is an intermediate point on a d_q-shortest permitted Ex_m-X path according to Definition 2.5 that is chosen such that it is l_2-visible from all points in C ($m = 1, \ldots, M$).

It follows immediately from Lemma 5.4 on page 64 that the functions $F_{q,X}(Y)$ are convex functions in $\mathbb{R}^2$. Therefore, Lemma 10.1 reveals that multiple objective location problems with barriers $1/P/(\mathcal{B} = N$ *convex polyhedra*)/d_B/$Q - f$ *convex* are closely related to the corresponding un-constrained problems $1/P/ \bullet /d/Q - f$ *convex*.

Observe also that the right-hand side of (10.7) takes on different values depending on what intermediate points have been used to evaluate the distance from a point X to the existing facilities. If these intermediate points are d_q-visible from another point $Y \in \mathcal{F}$, $q = 1, \ldots, Q$, then

$$F_{q,X}(Y) \geq F_{q,Y}(Y) \qquad \forall q = 1, \ldots, Q; \qquad (10.9)$$

cf. Lemma 5.5 on page 65. Inequality (10.9) holds, for example, for all points X, Y located in the same cell $C \in \mathcal{C}(\mathcal{G}_{l_2})$.

Consequently, a multiple objective location problem with barriers can be decomposed into a finite series of unconstrained multiple objective location problems within a cell of $\mathcal{C}(\mathcal{G}_{l_2})$ and with respect to the corresponding intermediate points.

Let the series of these subproblems be denoted by (P_k^i) $(i, k \in \mathbb{N})$ and have the following form:

$$\text{vmin} \quad \left(F_{1,k}^i(X), \ldots, F_{Q,k}^i(X)\right)^T \qquad (10.10)$$
$$\text{s.t.} \quad X \in \mathcal{F}^i,$$

where $\mathcal{F}^i = C^i$ is a cell of $\mathcal{C}(\mathcal{G}_{l_2})$ and the function $F_{q,k}^i$, $q \in \{1, \ldots, Q\}$, is defined according to (10.8), using the kth feasible assignment of intermediate points $I_{q,m,k}^i$ to existing facilities $Ex_m \in \mathcal{E}x$ with respect to the cell C^i. Thus, for $q = 1, \ldots, Q$ and $X \in \mathcal{F}^i$,

$$\begin{aligned}
F_{q,k}^i(X) &= f_q\Big(d_q(X, I_{q,1,k}^i) + d_{q,\mathcal{B}}(I_{q,1,k}^i, Ex_1), \ldots, \\
&\qquad d_q(X, I_{q,M,k}^i) + d_{q,\mathcal{B}}(I_{q,M,k}^i, Ex_M)\Big).
\end{aligned} \qquad (10.11)$$

In each subproblem (P_k^i), the intermediate points $I_{q,m,k}^i$, $q = 1, \ldots, Q$, $m = 1, \ldots, M$, are chosen such that they are l_2-visible from all points in the cell $\mathcal{F}^i = C^i$.

Observe that for all $X \in \mathcal{F}$ there exists a subproblem (P_k^i) such that $F_{q,X}(X) = F_{q,k}^i(X)$. Moreover, (10.9) can be rewritten as

$$F_{q,k}^i(Y) \geq F_{q,\bar{k}}^i(Y) \qquad \forall q = 1, \ldots, Q, \qquad (10.12)$$

where the subproblems (P_k^i) and $(P_{\bar{k}}^i)$ are chosen such that $F_{q,k}^i(X) = F_{q,X}(X)$ and $F_{q,\bar{k}}^i(Y) = F_{q,Y}(Y)$, and Y is an arbitrary point lying in the same cell as X.

The objective functions of problem (10.10) are convex functions, and the feasible set $\mathcal{F}^i$ of the subproblem (P_k^i) is equal to a cell of the grid $\mathcal{G}_{l_2}$ and is therefore a convex set for all values of i and k (see the discussion of the grid $\mathcal{G}_{l_2}$ on page 62). The number of subproblems (P_k^i) is finite, since only a finite number of possible combinations of intermediate points and existing facilities exists. Moreover, the definition of the grid $\mathcal{G}_{l_2}$ implies that $\bigcup_i \mathcal{F}^i = \mathcal{F}$.

Since the feasible sets $\mathcal{F}^i$ of the subproblems (10.10) are convex, every subproblem (P_k^i) is a restricted convex multiple objective problem for which connectedness of its efficient set is a well-known result from the literature (Warburton, 1983).

Let $\mathcal{X}_{E,k}^i$ and $\mathcal{Y}_{E,k}^i$ denote the sets of efficient solutions and of nondominated points of the problem (P_k^i), respectively. Individual minimization of each objective function $F_{q,k}^i(X)$ over the feasible set $\mathcal{F}^i$ produces the set of optimal solutions

$$\mathcal{X}_{q,k}^i := \arg \min_{X \in \mathcal{F}^i} \left\{ F_{q,k}^i(X) \right\}, \qquad q = 1, \ldots, Q,$$

and the optimal solution value

$$y_{q,k}^i = \min_{X \in \mathcal{F}^i} F_{q,k}^i(X), \qquad q = 1, \ldots, Q.$$

Having the efficient set of each convex subproblem available, we can specify their relationships with the efficient set of the nonconvex problem (10.4). Similarly, the nondominated set of this problem can be described by means of the nondominated set of the convex problems.

Theorem 10.1.

(i)
$$\mathcal{X}_{E,\mathcal{B}} \subseteq \bigcup_{i,k} \mathcal{X}_{E,k}^i$$

(ii)
$$\mathcal{Y}_{E,\mathcal{B}} = \mathrm{vmin} \bigcup_{i,k} \mathcal{Y}_{E,k}^i.$$

Proof.

(i) Let $X^* \in \mathcal{F}^i$ be an efficient solution of $1/P/(\mathcal{B} = N\,convex\ polyhedra)/d_\mathcal{B}/Q - f\,convex$. Lemma 10.1 and the definition of the subproblems (P_k^i) in (10.10) imply that there exists a $k \in \mathbb{N}$ such that

$$(f_{1,\mathcal{B}}(X^*), \ldots, f_{Q,\mathcal{B}}(X^*))^T = \left(F_{1,k}^i(X^*), \ldots, F_{Q,k}^i(X^*) \right)^T.$$

Assume that $X^* \notin \mathcal{X}_{E,k}^i$. Then there is a point $X^\circ \in \mathcal{X}_{E,k}^i$, $X^\circ \neq X^*$, such that

$$\left(F_{1,k}^i(X^\circ), \ldots, F_{Q,k}^i(X^\circ) \right)^T < \left(F_{1,k}^i(X^*), \ldots, F_{Q,k}^i(X^*) \right)^T.$$

Using the fact that $X^\circ \in \mathcal{F}^i$ and that therefore (10.7) and (10.12) can be applied for each objective $q = 1, \ldots, Q$, we obtain

$$(f_{1,\mathcal{B}}(X^\circ), \ldots, f_{Q,\mathcal{B}}(X^\circ))^T < (f_{1,\mathcal{B}}(X^*), \ldots, f_{Q,\mathcal{B}}(X^*))^T,$$

contradicting the fact that $X^* \in \mathcal{X}_{E,\mathcal{B}}$.

(ii) Part (ii) is a consequence of part (i) and the definition of efficient solutions.   □

Theorem 10.1 provides new information about the efficient sets and non-dominated sets of problem (10.4) and of the subproblems $\left(P_k^i\right)$, which will be used in the next section in the development of an algorithm for finding these sets in the biobjective case.

10.2   Methodology for Bicriteria Problems

In this section we study biobjective location problems with barriers, which we formulate as

$$\begin{aligned} \text{vmin} \quad & (f_{1,\mathcal{B}}(X), f_{2,\mathcal{B}}(X))^T \\ \text{s.t.} \quad & X \in \mathcal{F}, \end{aligned} \tag{10.13}$$

where $f_{1,\mathcal{B}}$ and $f_{2,\mathcal{B}}$ are defined according to (10.5).

Using Theorem 10.1, a straightforward algorithm to find the efficient set $\mathcal{X}_{E,\mathcal{B}}$ can be proposed. The algorithm first determines the set of efficient solutions of the corresponding subproblems $\left(P_k^i\right)$. Then, from the union of all the efficient sets $\mathcal{X}_{E,k}^i$ of the subproblems $\left(P_k^i\right)$, the efficient solutions of the original problem, referred to as *globally efficient solutions*, have to be determined. This can be done by constructing the *lower envelope* of all the nondominated points of the subproblems in the objective space.

Due to the definition of the subproblems $\left(P_k^i\right)$, their number depends upon the number of existing facilities and extreme points of the barrier regions. In general, it grows exponentially with the number of existing facilities if the number of feasible assignments of intermediate points to existing facilities cannot be limited. However, for example in the case of line barriers and two Weber objective functions based on the same metric $d_1 = d_2$, the consideration of a polynomial number of subproblems is sufficient if the subproblems are selected according to the selection procedure incorporated in Algorithms 7.1 and 7.2.

After the selection of an appropriate set of subproblems $\left(P_k^i\right)$ is completed, the resulting number of subproblems can be additionally reduced by applying a reduction procedure to eliminate subproblems whose nondominated sets are dominated by nondominated sets of other subproblems. We now discuss the details of this approach.

Let $List\left(P_k^i\right)$ be a list of all currently selected subproblems, containing $L := \left|List\left(P_k^i\right)\right|$ subproblems $\left(P_k^i\right)$. Since often only a small number of these subproblems contribute to the globally nondominated points, a reduction procedure is developed that reduces the number of subproblems before the globally nondominated points are finally determined as the lower envelope of the remaining sets $\mathcal{Y}_{E,k}^i$.

Consider a problem $\left(P_k^i\right)$ and its efficient and nondominated sets $\mathcal{X}_{E,k}^i$, $\mathcal{Y}_{E,k}^i$. Since $\left(P_k^i\right)$ is a convex problem. $\mathcal{Y}_{E,k}^i$ is a curve spanned between the points A_k^i and B_k^i, where

$$A_k^i = \left(a_{1,k}^i, a_{2,k}^i\right)^T \qquad \text{and} \qquad B_k^i = \left(b_{1,k}^i, b_{2,k}^i\right)^T$$

and

$$A_k^i = \operatorname*{lex\,min}_{X \in \mathcal{F}^i} \left\{F_{1,k}^i(X). F_{2,k}^i(X)\right\},$$
$$B_k^i = \operatorname*{lex\,min}_{X \in \mathcal{F}^i} \left\{F_{2,k}^i(X), F_{1,k}^i(X)\right\}.$$

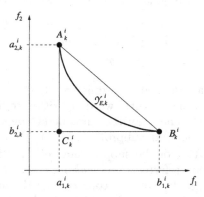

Figure 10.1. The nondominated set $\mathcal{Y}_{E,k}^i$ of a convex problem $\left(P_k^i\right)$.

As illustrated in Figure 10.1, the nondominated curve is contained in the triangle T_k^i with vertices A_k^i, B_k^i, C_k^i, where $C_k^i = \left(a_{1,k}^i, b_{2,k}^i\right)^T$. Observe that the examination of the respective locations of the triangles T_k^i will help eliminate those problems $\left(P_k^i\right)$ whose nondominated sets are dominated by nondominated sets of other subproblems.

Figure 10.2 shows four of many possible locations of the nondominated curves for two arbitrary problems $\left(P_k^i\right)$ and $\left(P_k^{\bar{i}}\right)$. In particular, Figure 10.2(a) shows that one of the two problems can be eliminated, while Figure 10.2(b) presents an irreducible case. Figures 10.2(c) and (d) show that only subsets of the two nondominated sets may be in the globally nondominated set. These observations will be incorporated into a reduction procedure as follows:

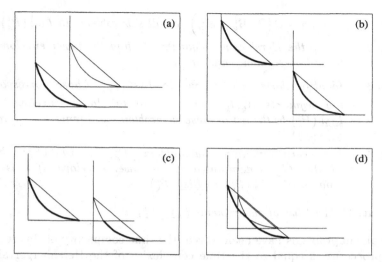

Figure 10.2. Some examples for possible locations of the triangles A_k^i, B_k^i, C_k^i and $A_k^{\bar{i}}, B_k^{\bar{i}}, C_k^{\bar{i}}$ for two different subproblems $\left(P_k^i\right)$ and $\left(P_k^{\bar{i}}\right)$ in the objective space. The bold curves represent the set of globally nondominated points, respectively.

In the first part of the procedure, the Hershberger algorithm (see Hershberger, 1989), which finds the lower envelope of a collection of line segments in linear time, is used to determine the lower envelope of the segments $\overline{A_k^i B_k^i}$ of all subproblems in $List\left(P_k^i\right)$. Since our goal is to find the nondominated sets of the subproblems, we add an auxiliary horizontal line at point B_k^i and an auxiliary vertical line at point A_k^i (this is equivalent to finding $\partial(\overline{A_k^i B_k^i} + \mathbb{R}_{\geq}^2)$) of each individual segment $\overline{A_k^i B_k^i}$ to eliminate points coming from other subproblems but dominated by the points of subproblem $\left(P_k^i\right)$.

After the lower envelope is found, all the subproblems contributing to it are selected and stored in a second list $\underline{List}\left(P_k^i\right)$.

In the second step of the procedure all those subproblems $\left(P_k^i\right)$ are added to the list $\underline{List}\left(P_k^i\right)$ for which at least one point (i.e., the point C_k^i) is not dominated by the lower envelope. Summarizing, the following procedure is obtained:

Algorithm 10.1. (Reduction Procedure)

Let $\mathbb{R}_{\geq}^2 := \left\{(x, y)^T \in \mathbb{R}^2 \; : \; x \geq 0, \; y \geq 0\right\}.$

Input: $List\left(P_k^i\right)$, *Segments* $\overline{A_k^i B_k^i}$.

Step 1:   Construct $\partial\left(\overline{A_k^i B_k^i} + \mathbb{R}_\geq^2\right)$ for all subproblems in List (P_k^i).

Step 2:   Apply the Hershberger algorithm to find the lower envelope of these line segments and half-lines.

Step 3:   Identify those subproblems in List (P_k^i) whose corresponding segments $\overline{A_k^i B_k^i}$ contribute to the lower envelope. Let $\underline{List}\,(P_k^i)$ be the list of these subproblems and remove them from List (P_k^i).

Step 4:   For every remaining subproblem $(P_k^i) \in$ List (P_k^i) check whether C_k^i is dominated by the lower envelope. If it is not dominated, add (P_k^i) to $\underline{List}\,(P_k^i)$.

Output: Reduced list of subproblems $\underline{List}\,(P_k^i)$.

Once all subproblems have been solved, the time complexity of this reduction procedure is equal to the time complexity of the Hershberger algorithm $O(r) = O(L \log L)$, where L denotes the number of subproblems in List (P_k^i). For many location problems the savings resulting from the reduction procedure will be substantial. However, they cannot be theoretically guaranteed.

The reduction procedure eliminates only subproblems (P_k^i) whose nondominated sets are entirely dominated by the nondominated set of other subproblems $\left(P_{\bar{k}}^{\bar{i}}\right)$ (see Figure 10.2(a)). Cases with partial reductions (see Figure 10.2(c),(d)) would be subject to further investigation.

Theorem 10.2.
$$\mathcal{Y}_{E,\mathcal{B}} = \text{vmin} \bigcup_{\underline{List}(P_k^i)} \mathcal{Y}_{E,k}^i.$$

Proof. Assume that there exists a subproblem $\left(P_{\bar{k}}^{\bar{i}}\right) \in List\,(P_k^i)$ such that $Y_{E,\bar{k}}^{\bar{i}}$ is globally nondominated, but $Y_{E,\bar{k}}^{\bar{i}} \notin \bigcup_{\underline{List}(P_k^i)} \mathcal{Y}_{E,k}^i$.

Since $Y_{E,\bar{k}}^{\bar{i}} \notin \bigcup_{\underline{List}(P_k^i)} \mathcal{Y}_{E,k}^i$, the corresponding point $C_{\bar{k}}^{\bar{i}}$ of the triangle $T_{\bar{k}}^{\bar{i}}$ of this subproblem is dominated by some point D in the lower envelope found by the Hershberger algorithm. Therefore, there exists a point $Y_{E,\hat{k}}^{\hat{i}} \in \bigcup_{\underline{List}(P_k^i)} \mathcal{Y}_{E,k}^i$, $Y_{E,\hat{k}}^{\hat{i}} \neq Y_{E,\bar{k}}^{\bar{i}}$, dominating D and thus dominating $Y_{E,\bar{k}}^{\bar{i}}$, which contradicts the assumption. This proves that $\mathcal{Y}_{E,\mathcal{B}} \subseteq \bigcup_{\underline{List}(P_k^i)} \mathcal{Y}_{E,k}^i$, which implies the desired result.   □

Recall that our ultimate goal is to determine the set of globally efficient solutions and globally nondominated points $\mathcal{X}_{E,\mathcal{B}}$ and $\mathcal{Y}_{E,\mathcal{B}}$ from the solutions $\mathcal{X}_{E,k}^i$ and $\mathcal{Y}_{E,k}^i$ of the individual subproblems. For this purpose the sets $\mathcal{X}_{E,k}^i$ and $\mathcal{Y}_{E,k}^i$ have to be found by available algorithms for the corresponding subproblems $(P_k^i) \in \underline{List}\,(P_k^i)$. Clearly, since biobjective location problems with barriers are generalizations of the corresponding unconstrained subproblems, we cannot expect to find better solution techniques for these problems than those known for the unconstrained problems.

We first discuss solution approaches for the case that each subproblem (P_k^i) involves only piecewise linear objective functions and its nondominated set $\mathcal{Y}_{E,k}^i$ is a piecewise linear curve. This is, for example, the case for Weber objective functions when distances are measured by the l_1 metric or by more general block norms. The problem (P_k^i) can be converted into a biobjective linear problem, and the parametric cost simplex method (see Geoffrion, 1967) can then be applied to exactly determine the nondominated set. Equivalently, the procedure of Nickel and Wiecek (1997) specially designed for biobjective piecewise linear programs can be used.

Given the nondominated sets of all the problems $(P_k^i) \in \underline{List}\,(P_k^i)$, we can determine the globally nondominated points, as proposed in Hamacher et al. (1999b), by means of the Hershberger algorithm (Hershberger, 1989). Since this algorithm finds a lower envelope of a collection of line segments, we again add an auxiliary horizontal line at point B_k^i and a vertical line at point A_k^i of every triangle T_k^i to eliminate points coming from other subproblems but dominated by the points of the subproblem (P_k^i). After the lower envelope has been found, these auxiliary lines are eliminated. The resulting lower envelope of the sets $\mathcal{Y}_{E,k}^i$ of all the subproblems $(P_k^i) \in \underline{List}\,(P_k^i)$ equals the set of globally nondominated points $\mathcal{Y}_{E,\mathcal{B}}$.

For other types of problems and other distance functions (such as the Weber problem with l_p-distance functions, $p \in (1,\infty)$) only approximation algorithms are known even in the unconstrained single-criterion case, and consequently the nondominated sets $\mathcal{Y}_{E,k}^i$ can be only approximated with a prescribed accuracy ε. The block sandwich method proposed by Yang and Goh (1997), which is based on the sandwich approximation of convex functions (see, for example, Burkard et al. (1991), Fruhwirth et al. (1989), and Rote (1992) for an overview), can produce piecewise linear upper and lower approximations of the nondominated sets. The method requires that one (approximately) solve scalarizations of the subproblems of the type

$$
\begin{aligned}
\min \quad & \lambda F_{1,k}^i(X) + (1-\lambda)F_{2,k}^i(X) \\
\text{s.t.} \quad & X \in \mathcal{F}^i
\end{aligned}
\tag{10.14}
$$

with weights $\lambda \in [0,1]$. A quadratic convergence property of this algorithm is established in Yang and Goh (1997); that is, the total number of optimization problems required to attain a prescribed approximation error is less than a constant multiple of the square root of the inverse of the given error.

The respective scalarized subproblems (10.14) can be solved by applying available methods for the corresponding single-criterion unconstrained location problems, as, for example, the Weiszfeld algorithm (Weiszfeld, 1937) in the case of Weber problems with l_p-distances, $p \in (1, \infty)$. In particular, the points A_i^k and B_i^k can be found by solving the corresponding single-objective location problems.

Alternatively, in order to approximate the nondominated sets of the subproblems for nonlinear objective functions one can use methods developed for general biobjective problems that produce discrete approximating sets of points (see Jahn and Merkel. 1992: Payne. 1993: Helbig, 1994). When connected, these points can become the input to the Hershberger algorithm (Hershberger, 1989). Since the nondominated sets are convex curves, one can also use the hyperellipse approach of Li et al. (1998), which is specially designed for convex biobjective problems. This approach produces a hyperellipse whose equation analytically represents the nondominated set. In this case, the Hershberger algorithm for curves in the plane (see Hershberger, 1989) can be applied, which can be viewed as a generalization of the above-mentioned algorithm for the construction of the lower envelope of a finite set of curves.

Whatever the method used to approximate the nondominated sets of the subproblems. these sets become the input to the Hershberger algorithm (Hershberger, 1989). since this algorithm finds a lower envelope of a collection of segments or of more general curves in the plane.

The discussion above leads to the following algorithm for solving biobjective location problems with barriers:

Algorithm 10.2. (Biobjective Location Problems with Barriers)

Input: *Location problem* $1/P/(\mathcal{B} = N$ *convex polyhedra*$)/d_{\mathcal{B}}/Q\text{--}f$ *convex.*

Step 1:   *Apply a selection procedure and create a list* $List\,(P_k^i)$ *of selected subproblems* (P_k^i).

Step 2:   *For every subproblem* $(P_k^i) \in List\,(P_k^i)$, *find the triangle* T_k^i.

Step 3:   *Apply the reduction procedure. Algorithm 10.1, and create a reduced list of subproblems* $\underline{List}\,(P_k^i)$.

Step 4:   *Construct the lower envelope of the sets* $\mathcal{Y}_{E,k}^i$ *corresponding to the subproblems* $(P_k^i) \in \underline{List}\,(P_k^i)$ *or of their (piecewise linear or convex) approximations determined with a prescribed error* ε.

Output: *The lower envelope is an exact representation of the set* $\mathcal{Y}_{E,\mathcal{B}}$ *or an approximation of the set* $\mathcal{Y}_{E,\mathcal{B}}$ *with error* ε, *depending on the available solution procedures for the corresponding unconstrained subproblems.*

Summarizing the discussion above. Algorithm 10.2 determines the nondominated set or an approximation of the nondominated set of the original problem, respectively, as the lower envelope of the nondominated sets of the subproblems, depending on the type of problem considered and on the given distance function. The complexity of the algorithm depends on the complexity of the methods used to solve the subproblems and on the required accuracy of the embedded approximation (cf. Yang and Goh, 1997). If the chosen method has polynomial complexity and if the number of selected subproblems L in $List\left(P_k^i\right)$ is polynomial (as, for example, in the case of Weber problems with line barriers; see Chapter 7 and Section 10.3 below), then Algorithm 10.2 is also polynomial.

Moreover, the proposed methodology produces solution approaches to bicriteria location problems with barriers as good as they can be for the corresponding single-criterion unconstrained problems. The algorithm gives exact solutions (i.e., finds all efficient solutions and nondominated points) for problems with piecewise linear objective functions whose nondominated set is piecewise linear but may be nonconvex.

More research is needed to design Algorithm 10.1 efficiently, eliminating some of the subproblems. Currently, the procedure checks only for the nondominated sets that are entirely dominated by nondominated sets of other subproblems. Cases with partial domination should also be considered.

Clearly, the complexity of more general location problems with barriers may heavily affect the ability to approximate their efficient sets. In this case, one may be interested in obtaining partial information about the set of efficient solutions and in designing tools for choosing a most preferred solution as the optimal one.

Furthermore, not only location problems can lead to nonconvex multiple objective problems decomposable to a series of convex problems. This class of nonconvex multiple objective problems should be explored independently of their applications.

10.3   An Example Problem with a Line Barrier

In the following example we consider a bicriteria Weber problem with a line barrier $1/P/\mathcal{B}_L/l_{1,\mathcal{B}_L}/2 - \sum$ where distances in both criteria are measured according to the Manhattan metric l_1.

For a given line barrier $\mathcal{B}_L$ (see Definition 7.1 on page 102) we denote by $\mathcal{F}^1$ and $\mathcal{F}^2$ the two closed half-planes on both sides of $\mathcal{B}_L$, and by $Ex_m^i \in \mathcal{F}^i$, $m \in \mathcal{M}^i := \{1, \ldots, M^i\}$, the set of existing facilities in each half-plane $\mathcal{F}^i$, $i = 1, 2$. A vector of nonnegative weights $w_{q,m}^i := w_q\left(Ex_m^i\right)$, $q = 1, 2$, is associated with each existing facility Ex_m^i representing the demand of Ex_m^i in the individual criterion. Thus the bicriteria Weber problem with a line barrier $1/P/\mathcal{B}_L/l_{1,\mathcal{B}_L}/2 - \sum$ can be formulated as

$$\text{vmin} \quad (f_{1,\mathcal{B}_L}(X), f_{2,\mathcal{B}_L}(X))^T$$
$$\text{s.t.} \quad X \in \mathcal{F} \tag{10.15}$$

with the individual objective functions given by

$$f_{q,\mathcal{B}_L}(X) = \sum_{i=1,2} \sum_{m=1}^{M^i} w_{q,m}^i \, l_{1,\mathcal{B}_L}\left(X, Ex_m^i\right), \qquad q = 1, 2. \qquad (10.16)$$

Since the same metric l_1 is assumed for both criteria, the assignment of intermediate points (i.e., passage points) to existing facilities is always the same for both criteria. Therefore, in the case of two passage points condition (7.6) on page 109 can be used to determine all the feasible assignments of intermediate points to existing facilities as described in Section 7.2. Analogously, condition (7.10) on page 112 can be used in the case of $N > 2$ passage points to determine the feasible assignments of intermediate points to existing facilities as described in Section 7.3.

In the case of two passage points, the resulting list of subproblems $List\left(P_k^i\right)$ contains at most $M^i + 1$ subproblems in the half-plane $\mathcal{F}^i$, $i = 1, 2$ (cf. Section 7.2, page 111), which we refer to as problems $\left(P_k^i\right)$, $i = 1, 2$ and $k = 0, 1, \ldots, M^i$, according to the respective half-plane $\mathcal{F}^i$ and the value of k in (7.7). In the case of $N > 2$ passage points we obtain an overall number of $O\left(\binom{M+N-1}{N-1}\right)$ subproblems, where $M = M^1 + M^2$ denotes the total number of existing facilities (cf. Section 7.3, page 115), which can be identified as discussed in Section 7.3.

Moreover, based on the results of Chapter 7, in each individual subproblem $\left(P_k^i\right)$ the restriction of the solutions to a cell of the visibility grid can be relaxed to a restriction to the corresponding half-plane $\mathcal{F}^i$. In particular, Theorem 10.1 combined with Theorems 7.2 and 7.1, and applied to problems with horizontal line barriers and the l_1-metric (recall the discussion of Theorem 7.2 on page 107) imply that the efficient set $\mathcal{X}_{E,\mathcal{B}_L}$ and the nondominated set $\mathcal{Y}_{E,\mathcal{B}_L}$ of $1/P/\mathcal{B}_L/l_{1,\mathcal{B}_L}/2 - \sum$ can be found as

$$\mathcal{X}_{E,\mathcal{B}_L} \subseteq \bigcup_{\substack{i=1,2;\\ k}} \mathcal{X}_{E,k}^i,$$

$$\mathcal{Y}_{E,\mathcal{B}_L} = \text{vmin} \bigcup_{\substack{i=1,2;\\ k}} \mathcal{Y}_{E,k}^i,$$

where $\mathcal{X}_{E,k}^i$ and $\mathcal{Y}_{E,k}^i$ denote the sets of efficient solutions and nondominated points of the relaxed subproblems $\left(P_k^i\right)$, respectively, and each subproblem $\left(P_k^i\right)$ is given by

$$\text{vmin} \quad \left(f_{1,k}^i(X) + g_{1,k}^j, f_{2,k}^i(X) + g_{2,k}^j\right)^T$$

$$\text{s.t.} \quad X \in \mathcal{F}^i$$

for $i, j \in \{1, 2\}$, $i \neq j$.

Here, $f_{q,k}^i$ and $g_{q,k}^i$ are defined according to (7.9) or (7.12), respectively, depending on the number of passage points. As an example, let the line barrier

$$\mathcal{B}_L := \left\{(x,y)^T \in \mathbb{R}^2 \; : \; y = 5\right\} \setminus \left\{P_1 = (4,5)^T, P_2 = (9,5)^T\right\}$$

divide the plane into the two half-planes $\mathcal{F}^1$ and $\mathcal{F}^2$ as shown in Figure 10.3. Furthermore, we assume that four existing facilities are given, two on each side of $\mathcal{B}_L$, with coordinates and weights as listed in Table 10.1. Thus $\mathcal{M}^1 = \mathcal{M}^2 = \{1,2\}$ and $M^1 = M^2 = 2$.

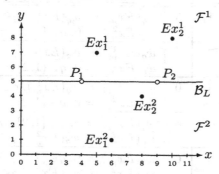

Figure 10.3. The example problem with the classification $1/P/\mathcal{B}_L/l_{1,\mathcal{B}_L}/2-\sum$.

Existing facility Ex_m^i		$w_{1,m}^i$	$w_{2,m}^i$	$D^i(m)$
Ex_1^1	$(5,7)^T$	8	2	-3
Ex_2^1	$(10,8)^T$	5	6	5
Ex_1^2	$(6,1)^T$	10	1	-1
Ex_2^2	$(8,4)^T$	7	4	3

Table 10.1. Existing facilities with their weights and the values of $D^i(m) = d\left(Ex_m^i, P_1\right) - d\left(Ex_m^i, P_2\right)$.

For the corresponding unconstrained Weber problems of type $1/P/\bullet/l_1/2-\sum$ exact algorithms are suggested in Hamacher and Nickel (1996). These algorithms are implemented in LoLA, the Library of Location Algorithms (see Hamacher et al., 1999a), which is used to find the exact efficient and nondominated sets of $1/P/\mathcal{B}_L/l_{1,\mathcal{B}_L}/2 - \sum$.

In Step 1 of Algorithm 10.2, six subproblems (P_k^i) are selected to be included in $List\left(P_k^i\right)$. These subproblems that are further investigated in the subsequent steps of Algorithm 10.2 are listed in Table 10.2. At the end of Step 3, the reduced list $\underline{List}\left(P_k^i\right)$ includes only the subproblems (P_0^1), (P_0^2), (P_1^2).

For illustrative reasons, we include the sets of efficient solutions (see Table 10.3) and nondominated points (see Figure 10.4) of all the subproblems in $List\left(P_k^i\right)$ which were determined using LoLA.

$\left(P_k^i\right)$	Weights of				Weights of existing facilities
	P_1		P_2		
	$\tilde{w}_1$	$\tilde{w}_2$	$\tilde{w}_1$	$\tilde{w}_2$	
$\left(P_0^1\right)$	0	0	17	5	$\tilde{w}_q(Ex_m^1) := w_q(Ex_m^1).\ q \in \{1,2\},\ m \in \mathcal{M}^1,$ $\tilde{w}_q(Ex_m^2) := 0.\ q \in \{1,2\},\ m \in \mathcal{M}^2$
$\left(P_1^1\right)$	10	1	7	4	$\tilde{w}_q(Ex_m^1) := w_q(Ex_m^1).\ q \in \{1,2\},\ m \in \mathcal{M}^1,$ $\tilde{w}_q(Ex_m^2) := 0.\ q \in \{1,2\}.\ m \in \mathcal{M}^2$
$\left(P_2^1\right)$	17	5	0	0	$\tilde{w}_q(Ex_m^1) := w_q(Ex_m^1).\ q \in \{1,2\},\ m \in \mathcal{M}^1,$ $\tilde{w}_q(Ex_m^2) := 0,\ q \in \{1,2\},\ m \in \mathcal{M}^2$
$\left(P_0^2\right)$	0	0	13	8	$\tilde{w}_q(Ex_m^1) := 0.\ q \in \{1,2\}.\ m \in \mathcal{M}^1.$ $\tilde{w}_q(Ex_m^2) := w_q(Ex_m^2).\ q \in \{1,2\},\ m \in \mathcal{M}^2$
$\left(P_1^2\right)$	8	2	5	6	$\tilde{w}_q(Ex_m^1) := 0.\ q \in \{1,2\}.\ m \in \mathcal{M}^1,$ $\tilde{w}_q(Ex_m^2) := w_q(Ex_m^2).\ q \in \{1,2\},\ m \in \mathcal{M}^2$
$\left(P_2^2\right)$	13	8	0	0	$\tilde{w}_q(Ex_m^1) := 0,\ q \in \{1,2\},\ m \in \mathcal{M}^1,$ $\tilde{w}_q(Ex_m^2) := w_q(Ex_m^2).\ q \in \{1,2\},\ m \in \mathcal{M}^2$

Table 10.2. Weights $\tilde{w}$ of the existing facilities $\mathcal{E}x = \{Ex_1^1, Ex_2^1, Ex_1^2, Ex_2^2, P_1, P_2\}$ of the six selected subproblems $\left(P_k^i\right)$ of type $1/P/\bullet/l_1/2 - \sum$.

Sub-problem	Efficient solutions of the subproblems $\mathcal{X}_{E,k}^i$
$\left(P_0^1\right)$	$\{(x,y)^T \in \mathbb{R}^2 : (x = 9) \wedge (5 \le y \le 7)\}$
$\left(P_1^1\right)$	$\{(x,y)^T \in \mathbb{R}^2 : (5 \le x \le 9) \wedge (y = 5)\}$ $\cup \{(x,y)^T \in \mathbb{R}^2 : (x = 9) \wedge (5 \le y \le 7)\}$
$\left(P_2^1\right)$	$\{(x,y)^T \in \mathbb{R}^2 : (4 \le x \le 5) \wedge (5 \le y \le 7)\}$
$\left(P_0^2\right)$	$\{(x,y)^T \in \mathbb{R}^2 : (8 \le x \le 9) \wedge (4 \le y \le 5)\}$
$\left(P_1^2\right)$	$\{(x,y)^T \in \mathbb{R}^2 : (6 \le x \le 8) \wedge (y = 4)\}$ $\cup \{(x,y)^T \in \mathbb{R}^2 : (x = 8) \wedge (4 \le y \le 5)\}$
$\left(P_2^2\right)$	$\{(x,y)^T \in \mathbb{R}^2 : (4 \le x \le 6) \wedge (4 \le y \le 5)\}$

Table 10.3. Efficient solutions of the six subproblems in $List\left(P_k^i\right)$.

Step 4 of Algorithm 10.2 yields the set of globally nondominated points $\mathcal{Y}_{E,\mathcal{B}_L}$, from which the set of globally efficient solutions $\mathcal{X}_{E,\mathcal{B}_L}$ of this example problem can be easily obtained:

$$
\begin{aligned}
\mathcal{Y}_{E,\mathcal{B}_L} &= \mathcal{Y}_{E,0}^1 \cup \mathcal{Y}_{E,0}^2 \cup \mathcal{Y}_{E,1}^2 \\
\mathcal{X}_{E,\mathcal{B}_L} &= \mathcal{X}_{E,0}^1 \cup \mathcal{X}_{E,0}^2 \cup \mathcal{X}_{E,1}^2 \\
&= \big\{ (x,y)^T \in \mathbb{R}^2 \ : \ ((x = 9) \wedge (5 \le y \le 7)) \\
&\qquad \vee ((8 \le x \le 9) \wedge (4 \le y \le 5)) \\
&\qquad \vee ((6 \le x \le 8) \wedge (y = 4)) \big\}.
\end{aligned}
$$

The set of globally efficient solutions $\mathcal{X}_{E,\mathcal{B}_L}$ is depicted in Figure 10.5.

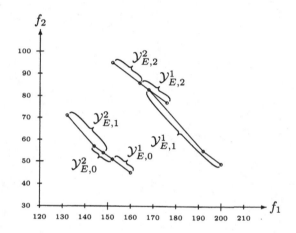

Figure 10.4. Nondominated points of the six subproblems in $List\,(P_k^i)$.

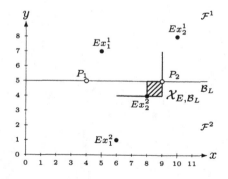

Figure 10.5. Efficient solutions $\mathcal{X}_{E,\mathcal{B}_L}$ of the example problem of type $1/P/\mathcal{B}_L/l_{1,\mathcal{B}_L}/2 - \sum$.

Part IV

Application

11
Location with Barriers Put to Work in Practice

City and regional planning is one of the most obvious areas for immediate application of planar location problems. Particularly in urban development, town planners are faced with a large variety of locational decisions. A new facility, for example, in a residential area, may be either a new fire department, a new hospital, a shopping center, or other public utility. Each of these requires a different type of objective function relating the already existing facilities with the new facilities.

This chapter deals with the particular case of locating a new playground into a developing residential area in the city of Halle (Saale), Germany. The problem was introduced in Kuhpfahl et al. (1995), where it was formulated as a multiple criteria location problem.

11.1 Problem Formulation

In a residential area of the city of Halle a new playground is to be located. The considered neighborhood includes a total of 18 apartment blocks. Each of these apartment blocks is identified by its coordinates with respect to a map of the area. The approximate number of children living in each of the respective units is provided by the city council. A rough map of the relevant part of Halle is given in Figure 11.1. Table 11.1 contains the coordinates of the existing apartment blocks together with the approximate number of children living in each of the buildings.

In Kuhpfahl et al. (1995) the problem of finding a "good" location for the planned playground is modeled as an unconstrained multicriteria planar location problem. Each of the 18 objective functions is defined by the approximated total walking distance from one of the buildings to the playground.

Coordinates of apartment blocks	Number of children
$(1.5, 5)^T$	58
$(2.5, 4)^T$	38
$(3, 5)^T$	58
$(3, 3)^T$	19
$(5, 3)^T$	115
$(5.5, 5.5)^T$	115
$(5.5, 8.5)^T$	58
$(6, 4)^T$	24
$(6, 3.5)^T$	24
$(6, 3)^T$	48
$(6.5, 4.5)^T$	19
$(6.5, 9)^T$	58
$(7, 3.5)^T$	38
$(9.5, 12.5)^T$	58
$(10, 4.5)^T$	77
$(10.5, 6)^T$	115
$(11, 12)^T$	77
$(11.5, 5.5)^T$	19

Table 11.1. The coordinates of the entrances to the 18 buildings and the approximate number of children living in each building.

To estimate walking distances in the considered part of Halle, the authors suggested the Manhattan metric l_1. since most of the streets in this neighborhood follow a rectilinear pattern. For both cases the sets of efficient solutions and nondominated points of the problem are determined, and a subsequent comparison of the resulting alternatives is performed.

In the following we will discuss a different approach to the problem that combines the individual objective functions of Kuhpfahl et al. (1995) by utilizing the information on the number of children living in each of the apartment blocks in order to identify a compromise objective function.

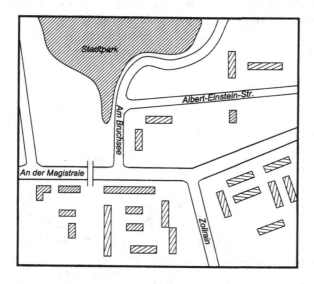

Figure 11.1. The considered part of the city of Halle, Germany.

Additionally, a regional constraint imposed by a highway intersecting the considered neighborhood is incorporated into the model. Only two bridges are available that allow a safe crossing of this highway for the children on their way to the playground. To account for the resulting detour on the way to the playground for some of the children, the highway is modeled by an almost linear barrier with two passage points at the location of the bridges.

The resulting model is illustrated in Figure 11.2, where the notation introduced in Chapter 7 is used to refer to the apartment buildings (the existing facilities) on the respective sides of the highway. The highway is interpreted as a piecewise linear barrier with the two passage points $P_1 = (4.0, 6.0)^T$ and $P_2 = (14.5, 8.5)^T$. Moreover, a park located in the area is included in the graph since it may be viewed as a forbidden region for the location of the playground. The dashed rectangle represents an area already belonging to the city and therefore preferred by the city council for the construction of the playground. A local train station is partially contained in this region, indicated by the small rectangle identified by the symbol S.

11.2 Mathematical Model: A Weber Problem with a Line Barrier

The structure of the problem suggests an application of the methods developed in Chapter 7. Even though the highway is only piecewise linear in reality, we will approximate it by a line while still working with the correct coordinates of the two passage points P_1 and P_2.

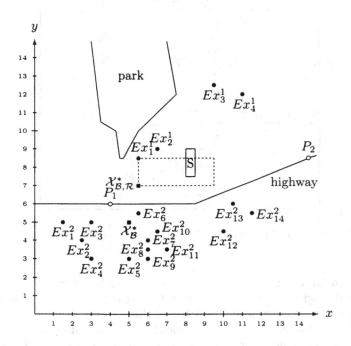

Figure 11.2. The existing apartment buildings and other relevant areas in the considered neighborhood of Halle.

Due to the special orientation of the streets in this part of the town (cf. Figure 11.1), the l_1 distance function seems to be suitable to approximate walking distances on each side of the dividing highway.

Since our goal is to find a location that minimizes the walking distance to the playground for as many children as possible, we use the Weber objective function with existing facilities at the locations of the apartment buildings, weighted by the approximate number of children living in each of the buildings. With the existing facilities as depicted in Figure 11.2, we obtain the following Weber problem with a line barrier:

$$\min \quad f_{\mathcal{B}_L}(X) = \sum_{m=1}^{M^1} w_m^1 l_{1,\mathcal{B}_L}\left(X, Ex_m^1\right) + \sum_{m=1}^{M^2} w_m^2 l_{1,\mathcal{B}_L}\left(X, Ex_m^2\right)$$
$$\text{s.t.} \quad X \in \mathcal{F},$$

$$(11.1)$$

where the feasible region $\mathcal{F}$ can be either the complete modeling horizon (i.e., the complete plane $\mathbb{R}^2$), or the rectangular area with the corner points $(5.5, 7.0)^T$, $(9.5, 7.0)^T$, $(9.5, 8.5)^T$, and $(5.5, 8.5)^T$ already owned by the city.

Existing facility Ex_m^i		$w_m^i = w\left(Ex_m^i\right)$	$D^i(m)$	$w_m^i \cdot l_1\left(Ex_m^i, P_1\right)$	$w_m^i \cdot l_1\left(Ex_m^i, P_2\right)$
Ex_1^1	$(5.5, 8.5)^T$	58	-5	232	522
Ex_2^1	$(6.5, 9)^T$	58	-3	319	493
Ex_3^1	$(9.5, 12.5)^T$	58	3	696	522
Ex_4^1	$(11, 12)^T$	77	6	1001	539
Ex_1^2	$(1.5, 5)^T$	58	-13.5	203	957
Ex_2^2	$(2.5, 4)^T$	38	-13	133	627
Ex_3^2	$(3, 5)^T$	58	-13	116	870
Ex_4^2	$(3, 3)^T$	19	-13	76	323
Ex_5^2	$(5, 3)^T$	115	-11	460	1725
Ex_6^2	$(5.5, 5.5)^T$	115	-10	230	1380
Ex_7^2	$(6, 4)^T$	24	-9	96	312
Ex_8^2	$(6, 3.5)^T$	24	-9	108	324
Ex_9^2	$(6, 3)^T$	48	-9	240	672
Ex_{10}^2	$(6.5, 4.5)^T$	19	-8	76	228
Ex_{11}^2	$(7, 3.5)^T$	38	-7	209	475
Ex_{12}^2	$(10, 4.5)^T$	77	-1	577.5	654.5
Ex_{13}^2	$(10.5, 6)^T$	115	0	747.5	747.5
Ex_{14}^2	$(11.5, 5.5)^T$	19	2	152	114

Table 11.2. Problem data for the Weber problem with a line barrier (11.1).

The coordinates of the entrances to the 18 existing facilities together with their weights, sorted according to the differences of distances to the two passage points P_1 and P_2 as defined in Section 7.2, and some additional problem data helpful for the determination of the objective function values of the resulting subproblems, are given in Table 11.2.

11.3 Solution

Algorithm 7.1 yields 20 unconstrained subproblems that are listed in Table 11.3. It can be easily verified that the minimum value of $F_k^i\left(X_k^i\right)$ is attained at the point $X_4^2 = (5.0, 5.0)^T$ with objective function value $f_{\mathcal{B}_L}\left((5.0, 5.0)^T\right)$ $= F_4^2\left(X_4^2\right) = 5350$. This implies that the unique optimal solution of (11.1) under the assumption that the complete modeling horizon is available for the new location is the point $X_{\mathcal{B}}^* = (5.0, 5.0)^T$; see Figure 11.2.

Sub-problem	Weights		Optimal solution of subproblem $\left(P_k^i\right)$			
$\left(P_k^i\right)$	$w(P_1)$	$w(P_2)$	$\mathcal{X}_k^i$	$f_k^i\left(X_k^i\right)$	g_k^j	$F_k^i\left(X_k^i\right)$
$\left(P_0^1\right)$	0	767	$(14.5, 8.5)^T$	2076	9409	11485
$\left(P_1^1\right)$	58	709	$(14.5, 8.5)^T$	2830	8655	11485
$\left(P_2^1\right)$	96	671	$(14.5, 8.5)^T$	3324	8161	11485
$\left(P_3^1\right)$	154	613	$(14.5, 8.5)^T$	4078	7407	11485
$\left(P_4^1\right)$	173	594	$(14.5, 8.5)^T$	4325	7160	11485
$\left(P_5^1\right)$	288	479	$(11.0, 8.5)^T$	5620	5895	11515
$\left(P_6^1\right)$	403	364	$(6.5, 8.5)^T$	6036	4745	10781
$\left(P_7^1\right)$	427	340	$(6.5, 8.5)^T$	5964	4529	10493
$\left(P_8^1\right)$	451	316	$[(5.5, 8.5)^T,$ $(6.0, 8.5)^T]$	5892	4313	10205
$\left(P_9^1\right)$	499	268	$(5.5, 8.5)^T$	5652	3881	9533
$\left(P_{10}^1\right)$	518	249	$(4.0, 6.0)^T$	5485	3729	9214
$\left(P_{11}^1\right)$	556	211	$(4.0, 6.0)^T$	4991	3463	8454
$\left(P_{12}^1\right)$	633	134	$(4.0, 6.0)^T$	3990	3386	7376
$\left(P_{13}^1\right)$	748	19	$(4.0, 6.0)^T$	2495	3386	5881
$\left(P_{14}^1\right)$	767	0	$(4.0, 6.0)^T$	2248	3424	5672

Sub-problem $\left(P_k^i\right)$	Weights		Optimal solution of subproblem $\left(P_k^i\right)$			
	$w(P_1)$	$w(P_2)$	$\mathcal{X}_k^i$	$f_k^i\left(X_k^i\right)$	g_k^j	$F_k^i\left(X_k^i\right)$
$\left(P_0^2\right)$	0	251	$(6.5, 5.0)^T$	5526	2076	7602
$\left(P_1^2\right)$	58	193	$(6.0, 5.0)^T$	5014	1786	6800
$\left(P_2^2\right)$	116	135	$(5.5, 5.0)^T$	4482	1612	6094
$\left(P_3^2\right)$	174	77	$(5.5, 5.0)^T$	3902	1786	5688
$\left(P_4^2\right)$	251	0	$(5.0, 5.0)^T$	3102	2248	5350

Table 11.3. The 20 unconstrained subproblems being solved while applying Algorithm 7.1. The solutions of the subproblems of the type $1/P/\bullet/l_1/\sum$ were obtained using the Library of Location Algorithms LoLA (see Hamacher et al., 1999a).

Observe that $X_{\mathcal{B}}^*$ is l_1-visible from all the existing facilities in the half-plane $\mathcal{F}^2$ and also from the two passage points P_1 and P_2. Therefore, the assumption that the barrier is a line in $\mathbb{R}^2$ (and not a piecewise linear curve) had no influence on the accuracy of the solution.

Alternatively, we can enforce a solution of (11.1) to be located in the rectangular feasible set $\left[(5.5, 7.0)^T, (9.5, 7.0)^T\right] \times \left[(5.5, 7.0)^T, (5.5, 8.5)^T\right]$. Thus only the 15 subproblems in the half-plane $\mathcal{F}^1$ are relevant in this case, where each of the individual subproblems is a restricted problem of the type $1/P/\mathcal{R}/l_1/\sum$. A list of the resulting subproblems in this case together with their optimal solutions and the optimal objective values is contained in Table 11.4.

The optimal solution of the restricted version of problem (11.1) lies at the point $\mathcal{X}_{\mathcal{B},\mathcal{R}}^* = (5.5, 7.0)^T$ and has an objective value of $f_{\mathcal{B}_L}\left((5.5, 7.0)^T\right) = F_{14}^1\left(X_{14}^1\right) = 6962$, see again Figure 11.2.

To compare the quality of different solutions of the problem (11.1), Figure 11.3 depicts the level sets $L_{\leq}(z, f_{\mathcal{B}}) = \{X \in \mathcal{F} : f_{\mathcal{B}}(X) \leq z\}$ of the objective function over the map of the neighborhood in question.

The nonconvexity of the objective function is nicely reflected in the nonconvexity of the level curves in Figure 11.3. Moreover, it can be noted that the passage point P_2 hardly influences the objective values, since it is used only to reach unfavorable solutions far from the center of the neighborhood.

An interesting open question is how the objective values could be improved by allowing for an additional passage point over the highway, and where such an additional bridge should be located in order to obtain a maximal improvement.

Sub-problem	Weights		Optimal solution of subproblem (P_k^i)			
(P_k^i)	$w(P_1)$	$w(P_2)$	$\mathcal{X}_k^i$	$f_k^i(X_k^i)$	g_k^j	$F_k^i(X_k^i)$
(P_0^1)	0	767	$(9.5, 8.5)^T$	4887	9409	14296
(P_1^1)	58	709	$(9.5. 8.5)^T$	5061	8655	13716
(P_2^1)	96	671	$(9.5. 8.5)^T$	5175	8161	13336
(P_3^1)	154	613	$(9.5. 8.5)^T$	5349	7407	12756
(P_4^1)	173	594	$(9.5. 8.5)^T$	5406	7160	15566
(P_5^1)	288	479	$(9.5. 8.5)^T$	5751	5895	11646
(P_6^1)	403	364	$(6.5. 8.5)^T$	6036	4745	10781
(P_7^1)	427	340	$(6.5. 8.5)^T$	5964	4529	10493
(P_8^1)	451	316	$[(5.5. 8.5)^T.$ $(6.0. 8.5)^T]$	5892	4313	10205
(P_9^1)	499	268	$(5.5, 8.5)^T$	5652	3881	9533
(P_{10}^1)	518	249	$(5.5, 7.0)^T$	5530	3729	9259
(P_{11}^1)	556	211	$(5.5. 7.0)^T$	5226	3463	8689
(P_{12}^1)	633	134	$(5.5. 7.0)^T$	4610	3386	7996
(P_{13}^1)	748	19	$(5.5. 7.0)^T$	3690	3386	7076
(P_{14}^1)	767	0	$(5.5. 7.0)^T$	3538	3424	6962

Table 11.4. The 15 subproblems of the type $1/P/\mathcal{R}/l_1/\sum$ being solved in the restricted case. The individual solutions of the subproblems were obtained using the appropriate routine implemented in the Library of Loaction Algorithms LoLA (Hamacher et al., 1999a).

11.4   Alternative Models

Instead of minimizing the overall walking distance of all the children to the playground one could also try to minimize the distance particularly to that apartment building that is at maximum distance from the new location in order to find a "fair" solution for all children using the playground. This approach yields a center problem with a line barrier that can be solved, for example, by the methods developed in Chapter 9, modeling the barrier by a suitable number of long and very thin rectangular sets.

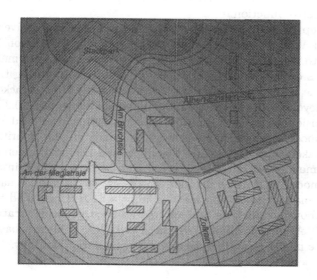

Figure 11.3. Level sets of the Weber problem (11.1) in the considered part of Halle.

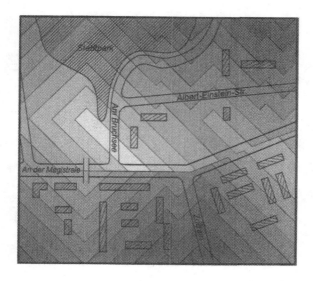

Figure 11.4. Level curves of the corresponding center problem in the considered neighborhood of Halle.

Figure 11.4 depicts the level curves of the problem with respect to the center objective function.

As a compromise between the Weber and the center objective functions an ordered Weber objective function as formulated in Section 3.1 could be chosen to model the problem. Other alternatives include a multicriteria approach as discussed in Kuhpfahl et al. (1995) incorporating the barrier given by the highway. In this case the problem could be attacked by the methods developed in Chapter 10; see in particular Section 10.3.

The decision for a most suitable problem formulation and the corresponding modeling parameters strongly depends on the priorities of the decision-makers and often includes political and personal preferences. Obviously, some of these criteria are suitable to be incorporated into a mathematical model, and other decision processes, by their very nature, will deny their mathematical treatment. The present text has tried to give an overview about possible solution approaches once a specific location model is selected. A good locational decision, however, is always based on an appropriate model as well as the appropriate solution strategy.

References

Alt, H. and Welzl, E. (1988). Visibility graphs and obstacle-avoiding shortest paths. *Zeitschrift für Operations Research*, **32**, 145–164.

Alt, H. and Yap, C.-K. (1990). Algorithmic aspects of motion planning: A tutorial. *Algorithms Review*, **1**, 43–77.

Aneja, Y.P. and Parlar, M. (1994). Algorithms for Weber facility location in the presence of forbidden regions and/or barriers to travel. *Transportation Science*, **28**, 70–76.

Armacost, R.L. and Mylander, W.C. (1973). A guide to SUMT – version 4. Computer subroutine for implementing sensitivity analysis in nonlinear programming. Technical Report T-287, The Institute of Management Science and Engineering, The George Washington University, D.C.

Batta, R. and Leifer, L.A. (1988). On the accuracy of demand point solutions to the planar, Manhattan metric, p-median problem, with and without barriers to travel. *Computers and Operations Research*, **15**, 253–262.

Batta, R., Ghose, A., and Palekar, U.S. (1989). Locating facilities on the Manhatten metric with arbitrarily shaped barriers and convex forbidden regions. *Transportation Science*, **23**, 26–36.

Bazaraa, M.S., Sherali, H.D., and Shetty, C.M. (1993). *Nonlinear Programming*. Wiley, New York.

Ben-Moshe, B., Katz, M.J., and Mitchell, J.S.B. (2001). Farthest neighbors and center points in the presence of rectangular obstacles. In *Proc. 17th ACM Symposium on Computational Geometry*.

Bentley, J.L. and Ottmann, T.A. (1979). Algorithms for reporting and counting geometric intersections. *IEEE Trans. Comput.*, **C-28**, 643–647.

Berens, W. (1988). The suitability of the weighted l_p-norm in estimating actual road distances. *European Journal of Operational Research*, **34**, 39–43.

Berens, W. and Koerling, F.J. (1985). Estimating road distances by mathematical functions. *European Journal of Operational Research*, **21**, 54–56.

Berens, W. and Koerling, F.J. (1988). On estimating road distances by mathematical functions – a rejoinder. *European Journal of Operational Research*, **36**, 254–255.

Berman, O., Larson, R.C., and Chiu, S.S. (1985). Optimal server location on a network operating as an $m/g/1$ queue. *Operations Research*, **33**, 746–771.

Bitran, G.R. and Magnanti, T.L. (1979). The structure of admissible points with respect to cone dominance. *Journal of Optimization Theory and Applications*, **29**, 573–614.

Brady, S.D. and Rosenthal, R. (1980). Interactive computer graphical solutions to constrained minimax location problems. *AIIE Transactions*, **12**, 241–248.

Brady, S.D. and Rosenthal, R. (1983). Interactive graphical minimax location of multiple facilities with general constraints. *IIE Transactions*, **15**, 242–252.

Brimberg, J. and Love, R.F. (1991). Estimating travel distances by the weighted l_p norm. *Naval Research Logistics*, **38**, 241–259.

Brimberg, J. and Love, R.F. (1992). A new distance function for modeling travel distances in a transportation network. *Transportation Science*, **26**, 129–137.

Brimberg, J. and Love, R.F. (1995). Estimating distances. In Z. Drezner, editor, *Facility Location*, pages 225–262. Springer Series in Operations Research.

Brimberg, J., Dowling, P.D., and Love, R.F. (1996). Estimating the parameters of the weighted l_p norm by linear regression. *IIE Transactions*, **28**, 363–367.

Bronstein, I.N. and Semendjajew, K.A. (1985). *Taschenbuch der Mathematik*. Teubner, Leipzig. 22nd. edition.

Burkard, R.E., Hamacher, H.W., and Rote, G. (1991). Sandwich approximation of univariate convex functions with an application to separable convex programming. *Naval Research Logistics*, **38**, 911–924.

Butt, E.S. (1994). *Facility Location in the Presence of Forbidden Regions*. Ph.D. thesis, Department of Industrial and Management Systems Engineering, Pennsylvania State University, PR.

Butt, S.E. and Cavalier, T.M. (1996). An efficient algorithm for facility location in the presence of forbidden regions. *European Journal of Operational Research*, **90**, 56–70.

Butt, S.E. and Cavalier, T.M. (1997). Facility location in the presence of congested regions with the rectilinear distance metric. *Socio-Economic Planning Sciences*, **31**, 103–113.

Chalmet, L., Francis, R., and Kolen, A. (1981). Finding efficient solutions for rectilinear distance location problems efficiently. *European Journal of Operational Research*, **6**, 117–124.

Chazelle, B. and Edelsbrunner, H. (1992). An optimal algorithm for intersecting line segments in the plane. *J. Assoc. Comput. Mach.*, **39**, 1–54.

Chiu, S.S., Berman, O., and Larson, R.C. (1985). Locating a mobile server queueing facility on a tree network. *Management Science*, **31**, 764–772.

Choi, J., Sellen, J., and Yap, C.-K. (1995). Precision-sensitive Euclidean shortest path in 3-space. In *Proceedings of the 11th Annual ACM Symposium on Computational Geometry (SCG)*, pages 350–359.

Choi, J., Shin, C.-S., and Kim, S.K. (1998). Computing weighted rectilinear median and center set in the presence of obstacles. In *Proc. 9th Annu. Internat. Sympos. Algorithms Comput.*, pages 29–38. Vol. 1533 of *Lecture Notes in Computer Science*, Springer-Verlag.

Clarke, F.H. (1983). *Optimization and Nonsmooth Analysis*. Wiley, New York.

Clarkson, K.L. and Shor, P.W. (1989). Applications of random sampling in computational geometry, II. *Discrete Computational Geometry*, **4**, 387–421.

Daskin, M.S. (1995). *Network and Discrete Location*. Wiley, New York.

De Berg, M., Matoušek, J., and Schwarzkopf, O. (1995). Piecewise linear paths among convex obstacles. *Discrete and Computational Geometry*, **14**, 9–29.

Dearing, P.M. and Segars Jr., R. (2000a). An equivalence result for planar location problems with rectilinear distance and barriers. Technical report, Department of Mathematical Sciences, Clemson University, SC.

Dearing, P.M. and Segars Jr., R. (2000b). Solving rectilinear planar location problems with barriers by a polynomial partitioning. Technical report, Department of Mathematical Sciences, Clemson University, SC.

Dearing, P.M., Hamacher, H.W., and Klamroth, K. (2002). Center problems with barriers and the Manhattan metric. *Naval Research Logistics*. To appear.

Ding, Y., Baveja, A., and Batta, R. (1994). Implementing Larson and Sadiq's location model in a geographic information system. *Computers and Operations Research*, **21**, 447–454.

Drezner, Z. (1983). Constrained location problems in the plane and on a sphere. *IEE Transactions*, **12**, 300–304.

Drezner, Z., editor (1995). *Facilities Location: A Survey of Applications and Methods*. Springer-Verlag, New York.

Drezner, Z. and Goldman, A.J. (1991). On the set of optimal points to the Weber problem. *Transportation Science*, **25**, 3–8.

Drezner, Z. and Hamacher, H.W., editors (2002). *Facility Location: Applications and Theory*. Springer-Verlag, New York.

Drezner, Z., Klamroth, K., Schöbel, A., and Wesolowsky, G.O. (2002). The Weber problem. In Z. Drezner and H.W. Hamacher, editors, *Facility Location: Applications and Theory*, pages 1–36. Springer-Verlag, New York.

Durier, R. (1990). On Pareto optima, the Fermat-Weber problem, and polyhedral gauges. *Mathematical Programming*, **47**, 65–79.

Durier, R. and Michelot, C. (1985). Geometrical properties of the Fermat-Weber Problem. *Journal of Mathematical Analysis and Applications*, **117**, 506–528.

Durier, R. and Michelot, C. (1994). On the set of optimal points to the Weber problem: Further results. *Transportation Science*, **28**, 141–149.

Eckhardt, U. (1975). On an optimization problem related to minimal surfaces with obstacles. In *Optimization and Optimal Control, Lecture Notes in Mathematics, Vol. 477*. Springer-Verlag, Berlin.

Edelsbrunner, H. (1987). *Algorithms in Combinatorial Geometry.* Springer-Verlag, Berlin.

Ehrgott, M. (2000). *Multicriteria Optimization.* Lecture Notes in Economics and Mathematical Systems 491, Springer-Verlag, Berlin.

Ehrgott, M., Nickel, S., and Hamacher, H.W. (1999). Geometric methods to solve max-ordering location problems. *Discrete Applied Mathematics*, **93**, 3–20.

Elsgolc, L.E. (1962). *Calculus of Variations.* International Series of Monographs in Pure and Applied Mathematics, Vol. 19, Addison Wesley.

Fekete, S.P., Mitchell, J.S.B., and Weinbrecht, K. (2000). On the continuous Weber and k-median problems. Technical Report No. 666/2000, Technische Universität Berlin, Department of Mathematics.

Fernández, J., Cánovas, L., and Pelegrín, B. (1997). DECOPOL – Codes for decomposing a polygon into convex subpolygons. *European Journal of Operational Research (O.R.S.E.P. Section)*, **102**, 242–243.

Fernández, J., Cánovas, L., and Pelegrín, B. (2000). Algorithms for the decomposition of a polygon into convex polygons. *European Journal of Operational Research*, **121**, 330–342.

Fernández, J., Fernández, P., and Pelegrín, B. (2002). Estimating actual distances by norm functions: A comparison between the $l_{k,p,\theta}$-norm and the $l_{b_1,b_2,\theta}$-norm and a study about the selection of the data set. *Computers and Operations Research*, **29**, 609–623.

Fiacco, A.V. and McCormic, G.V. (1968). *Nonlinear Programming: Sequential Unconstrained Minimization Techniques.* Wiley, New York.

Fliege, J. (1997). *Effiziente Dimensionsreduktion in Multilokationsproblemen.* Shaker Verlag, Aachen.

Fliege, J. (2000). Solving convex location problems with gauges in polynomial time. *Studies in Locational Analysis*, **14**, 153–172.

Fliege, J. and Nickel, S. (1997). An interior point method for multifacility location problems with forbidden regions. Technical Report in Angewandter Mathematik No. 142, Universität Dortmund, Department of Mathematics.

Forster, O. (1977). *Analysis 2. Differentialrechnung im $\mathbb{R}^n$; Gewöhnlich Differentialgleichungen.* Vieweg Verlag, Braunschweig.

Forster, O. (1983). *Analysis 1. Differential- und Integralrechnung eine Veränderlichen.* Vieweg Verlag, Braunschweig.

Foulds, L.R. and Hamacher, H.W. (1993). Optimal bin location in printed circuit board assembly. *European Journal of Operational Research*, **66**, 279–290.

Francis, R.L., Leon, F., McGinnis, L.F.. and White, J.A. (1992). *Facility Layout and Location: An Analytical Approach*. Prentice-Hall, New York, 2nd edition.

Frieß, L. (1999). *Lösung planarer Centerprobleme mit Hilfe physikalischer Wellenmodelle – Computergestützte Simulation der Lösungsmethode für die l_1-Metrik*. Staatsexamensarbeit am Fachbereich Mathematik der Universität Kaiserslautern.

Fruhwirth, B., Burkard, R.E., and Rote, G. (1989). Approximation of convex curves with application to the bi-criteria minimum cost flow problem. *European Journal of Operational Research*, **42**, 326–338.

Geoffrion, A.M. (1967). Solving bicriterion mathematical programs. *Operations Research*, **15**, 39–54.

Gerth, C. and Pöhler, K. (1988). Dualität und algorithmische Anwendung beim vektoriellen Standortproblem. *Optimization*, **19**, 491–512.

Ghosh, S.K. and Mount, D.M. (1991). An output-sensitive algorithm for computing visibility graphs. *SIAM Journal on Computing*, **20**, 888–910.

Goldman, A.J. (1971). Optimal center location in simple networks. *Transportation Science*, **5**, 121–221.

Hamacher, H.W. (1995). *Mathematische Lösungsverfahren für planare Standortprobleme*. Vieweg Verlag, Braunschweig.

Hamacher, H.W. and Klamroth, K. (2000). Planar location problems with barriers under polyhedral gauges. *Annals of Operations Research*, **96**, 191–208.

Hamacher, H.W. and Nickel, S. (1993). Median location problems with several objectives. *Studies in Locational Analysis*, **4**, 149–153.

Hamacher, H.W. and Nickel, S. (1994). Combinatorial algorithms for some 1-facility median problems in the plane. *European Journal of Operational Research*, **79**, 340–351.

Hamacher, H.W. and Nickel, S. (1995). Restricted planar location problems and applications. *Naval Research Logistics*, **42**, 967–992.

Hamacher, H.W. and Nickel, S. (1996). Multicriteria planar location problems. *European Journal of Operational Research*, **94**, 66–86.

Hamacher, H.W. and Nickel, S. (1998). Classification of location problems. *Location Science*, **6**, 229–242.

Hamacher, H.W. and Nickel, S. (2001). Multi-facility and restricted location problems. In C.A. Floudas and P.M. Pardalos, editors, *Encyclopedia of Optimization*. Kluwer Academic Publishers, Dordrecht.

Hamacher, H.W. and Schöbel, A. (1997). A note on center problems with forbidden polyhedra. *Operations Research Letters*, **20**, 165–169.

Hamacher, H.W., Hennes, H., Klamroth, K., Müller, M.C., Nickel, S., and Schöbel, A. (1999a). LoLA: Library of Location Algorithms – a toolkit for solving location problems. http://www.mathematik.uni-kl.de/~lola/. Software of the University of Kaiserslautern, Germany, Release 2.0.

Hamacher, H.W., Labbé, M., and Nickel, S. (1999b). Multicriteria network location problems with sum objectives. *Networks*, **33**, 79–92.

Handler, G.Y. and Mirchandani, P.B. (1979). *Location on Networks*. MIT Press, Cambridge, MA.

Hansen, P., Perreur, J., and Thisse, J.-F. (1980). Location theory, dominance and convexity: Some further results. *Operations Research*, **28**, 1241–1250.

Hansen, P., Peeters, D., and Thisse, J.-F. (1982). An algorithm for a constrained Weber problem. *Management Science*, **28**, 1285–1295.

Hansen, P., Peeters, D., Richard, D., and Thisse, J.-F. (1985). The minisum and minimax location problems revisited. *Operations Research*, **33**, 1251–1265.

Hansen, P., Jaumard, B., and Tuy, H. (1995). Global optimization in location. In Z. Drezner, editor, *Facility Location*, pages 43–68. Springer Series in Operations Research.

Hansen, P., Krau, S., Peeters, D., and Thisse, J.-F. (2000). Weber's problem with forbidden regions for location and transportation. Technical Report G-2000-26, Les Cahiers du Gerad.

Harary, F. (1969). *Graph Theory*. Addison-Wesley, Reading, MA.

Helbig, S. (1994). On a constructive approximation of the efficient outcomes in bicriterion vector optimization. *Journal of Global Optimization*, **5**, 35–48.

Hershberger, J. (1989). Finding the upper envelope of n line segments in $O(n \log n)$ time. *Information Processing Letters*, **33**, 169–174.

Hooke, R. and Jeeves, T.A. (1961). Direct search solution of numerical and statistical problems. *J. Assoc. Comput. Mach.*, **8**, 212–229.

Icking, C., Klein, R., Ma, L., Nickel, S., and Weißler, A. (1999). On bisectors for different distance functions. In *Fifteenth Annual ACM Symposium on Computatational Geometry*.

Jahn, J. and Merkel, A. (1992). Reference point approximation method for the solution of bicriteria nonlinear optimization problems. *Journal of Optimization Theory and Applications*, **74**, 87–103.

Juel, H. and Love, R.F. (1983). Hull properties in location problems. *European Journal of Operational Research*. **12**, 262–265.

Kaliszewski, I. (1994). *Quantitative Pareto Analysis by Cone Separation Technique*. Kluwer Academic Publisher, Dordrecht.

Karkazis, J. (1988). The general unweighted problem of locating obnoxious facilities on the plane. *Belgian Journal of Operations Research, Statistics and Computer Science*, **28**, 3–49.

Katz, I.N. and Cooper, L. (1979a). Facility location in the presence of forbidden regions, II: Euclidean distance and several forbidden circles. Technical Report OREM 79006. Southern Methodist University, Dallas, TX 75275.

Katz, I.N. and Cooper, L. (1979b). Facility location in the presence of forbidden regions, III: l_p distance and polygonal forbidden regions. Technical Report OREM 79011. Southern Methodist University, Dallas, TX 75275.

Katz, I.N. and Cooper, L. (1981). Facility location in the presence of forbidden regions, I: Formulation and the case of Euclidean distance with one forbidden circle. *European Journal of Operational Research*, **6**, 166–173.

Kirk, D. and Lim, L. (1970). A dual-mode routing algorithm for an autonomous moving vehicle. *IEEE Transactions on Aerospace and Electronic Systems*, **6**, 290–294.

Klamroth, K. (2001a). Planar Weber location problems with line barriers. *Optimization*, **49**, 517–527.

Klamroth, K. (2001b). A reduction result for location problems with polyhedral barriers. *European Journal of Operational Research*, **130**, 486–497.

Klamroth, K. and Wiecek, M.M. (2002). A bi-objective median location problem with a line barrier. *Operations Research*. To appear.

Krau, S. (1996). *Extensions du Problème de Weber.* Ph.D. thesis, Département de Mathématiques et de Génie Industriel, Université de Montréal.

Kuhpfahl, I., Patz, R., and Tammer, Chr. (1995). Location problems in town planning. In D. Schweigert, editor, *Methods of Multicriteria Decision Theory, Proc. of the 5th Workshop of DGOR-Working Group Multicriteria Optimization and Decision Theory*, pages 101–112.

Kusakari, Y. and Nishizeki, T. (1997). An algorithm for finding a region with the minimum total L_1 from prescribed terminals. In *Proc. of ISAAC'97.* Vol. 1350 of *Lecture Notes in Computer Science*, Springer-Verlag.

LaPaugh, A.S. (1980). *Algorithms for Integrated Circuit Layout: An Analytic Approach.* Ph.D. thesis, Massachusetts Institute of Technology.

Larson, R.C. and Li, V.O.K. (1981). Finding minimum rectilinear distance paths in the presence of barriers. *Networks*, **11**, 285–304.

Larson, R.C. and Sadiq, G. (1983). Facility locations with the Manhattan metric in the presence of barriers to travel. *Operations Research*, **31**, 652–669.

Latombe, J.-C. (1991). *Robot Motion Planning.* Kluwer Academic Publishers, Dordrecht.

Li, Y., Fadel, G., and Wiecek, M.M. (1998). Approximating Pareto curves using the hyper-ellipse. In *Proceedings of the 7th AIAA/ USAF/NASA/ISSMO Multidisciplinary Analysis and Optimization Symposium*, volume 3, pages 1990–2002.

Love, R.F. and Morris, J.G. (1972). Modelling inter-city road distances by mathematical functions. *Operational Research Quaterly*, **23**, 61–71.

Love, R.F. and Morris, J.G. (1975). A computation procedure for the exact solution of location-allocation problems with rectangular distances. *Naval Research Logistics Quaterly*, **22**, 441–453.

Love, R.F. and Morris, J.G. (1979). Mathematical models of road travel distances. *Management Science*, **25**, 130–139.

Love, R.F. and Morris, J.G. (1988). On estimating road distances by mathematical functions. *European Journal of Operational Research*, **36**, 251–253.

Love, R.F. and Walker, J.H. (1994). An empirical comparison of block and round norms for modelling actual distances. *Location Science*, **2**, 21–43.

Love, R.F., Morris, J.G., and Wesolowsky, G.O. (1988). *Facilities Location: Models & Methods*. North Holland, New York.

Lozano-Perez, T. and Wesley, M. (1979). An algorithm for planning collision free paths among polyhedral obstacles. *Communications of the ACM*, **22**, 560–570.

Luc, D.T. (1989). *Theory of Vector Optimization*. Lecture Notes in Economics and Mathematical Systems 319. Springer-Verlag, Berlin.

McGinnis, L. and White, J. (1978). A single facility rectilinear location problem with multiple criteria. *Transportation Science*, **12**, 217–231.

Minkowski, H. (1911). *Gesammelte Abhandlungen, zweiter Band*. Editor: D. Hilbert. Teubner Verlag, Leipzig und Berlin. Also in: Chelsea Publishing Company, New York. 1967.

Mirchandani, P.B. and Francis, R.L. (1990). *Discrete Location Theory*. Wiley, New York.

Mitchell, J.S.B. (1992). L_1 shortest paths among polygonal obstacles in the plane. *Algorithmica*, **8**. 55–88.

Mitchell, J.S.B. (2000). Geometric shortest paths and network optimization. In J.-R. Sack and J. Urrutia, editors. *Handbook of Computational Geometry*, pages 633–701. North-Holland, Amsterdam.

Mitchell, J.S.B., Rote, G., and Woeginger, G. (1992). Minimum-link paths among obstacles in the plane. *Algorithmica*, **8**, 431–459.

Müller, A. (1997). *Polyedrische Gauges als Approximation für Netzwerkentfernungen und ihre Anwendung für Geographische Informationssysteme (GIS)*. Diplomarbeit am Fachbereich Mathematik der Universität Kaiserslautern.

Nickel, S. (1991). *Restriktive Standortprobleme*. Diplomarbeit am Fachbereich Mathematik der Universität Kaiserslautern.

Nickel, S. (1994). Multicriterial and restricted location problems with polyhedral gauges. In *Operations Research Proceedings*, pages 109–114. Springer-Verlag, Berlin.

Nickel, S. (1995). *Discretization of Planar Location Problems*. Shaker Verlag, Aachen.

Nickel, S. (1997). Bicriteria and restricted 2-facility Weber problems. *Mathematical Methods of Operations Research*, **45**, 167–195.

Nickel, S. (1998). Restricted center problems under polyhedral gauges. *European Journal of Operational Research*, **104**, 343–357.

Nickel, S. and Hamacher, H.W. (1992). RLP: A program package for solving restricted 1-facility location problems in a user friendly environment. *European Journal of Operational Research*, **62**, 116–117.

Nickel, S. and Puerto, J. (1999). A unified approach to network location problems. *Networks*, **34**, 283–290.

Nickel, S. and Wiecek, M.M. (1997). A flexible approach to piecewise linear multiple objective programming. In U. Zimmermann, U. Derigs, W. Gaul, Möhring R.H., and K.-P. Schuster, editors, *Operations Research Proceedings 1996*. Springer-Verlag, Berlin.

Ochs, M. (1998). *1-Standort Medianprobleme mit einfachen Barrieren*. Diplomarbeit, Department of Mathematics, University of Kaiserslautern, Germany.

Okabe, A., Boots, B., and Sugihara, K. (1992). *Spatial Tesselations: Concepts and Applications of Voronoi Diagrams*. Wiley, Chichester.

O'Rourke, J. (1993). Computational geometry column 18. *International Journal on Computational Geometry and Applications*, **2**, 107–113. Also in: *SIGACT News*, **24**, 20-25, 1993.

O'Rourke, J. (1994). *Computational Geometry in C*. Cambridge University Press.

Overmars, M.H. and Welzl, E. (1988). New methods for computing visibility graphs. In *Proceedings of the 4th Annual ACM Symposium on Computational Geometry (SCG)*, pages 164–171.

Payne, A.N. (1993). Efficient approximate representation of bi-objective tradeoff sets. *Journal of the Franklin Institute*, **330**, 1219–1233.

Pelegrin, P., Michelot, C., and Plastria, F. (1985). On the uniqueness of optimal solutions in continuous location theory. *European Journal of Operational Research*, **20**, 327–331.

Perreur, J. (1989). L'évolution des représentations de la distance et l'aménagement du territoire. *Revue D'Economie Régionale et Urbaine*, **1**, 5–32.

Plastria, F. (1984). Localization in single facility location. *European Journal of Operational Research*, **18**, 215–219.

Plastria, F. (1992). GBSSS, the Generalized Big Square Small Square method for planar single facility location. *European Journal of Operational Research*, **62**, 163–174.

Plastria, F. (1995). Continuous location problems. In Z. Drezner, editor, *Facility Location*, pages 225–262. Springer Series in Operations Research.

Plastria, F. (2002). Continuous covering location problems. In Z. Drezner and H.W. Hamacher, editors, *Facility Location: Applications and Theory*, pages 37–79. Springer-Verlag, New York.

Polak, E. (1997). *Optimization – Algorithms and Consistent Approximations*. Springer-Verlag, New York.

Puerto, J. and Fernández, F.R. (2000). Geometrical properties of the symmetrical single facility location problem. *Journal of Nonlinear and Convex Analysis*, **1**, 321–342.

Rockafellar, R.T. (1970). *Convex Analysis*. Princeton University Press.

Rockafellar, R.T. and Wets, R.J.-B. (1998). *Variational Analysis*. Springer-Verlag, Berlin.

Rodriguez-Chia, A.M., Nickel, S., Puerto, J., and Fernandez, F.R. (2000). A flexible approach to location problems. *Mathematical Methods of Operations Research*, **51**, 69–89.

Roos, T. and Noltemeier, H. (1991). Dynamic Voronoi diagrams in motion planning. In H. Bieri and H. Noltemeier, editors, *Computational Geometry – Methods, Algorithms and Applications*, pages 227–236. Vol. 553 of *Lecture Notes in Computer Science*, Springer-Verlag.

Rote, G. (1992). The convergence rate of the sandwich algorithm for approximating convex functions. *Computing*, **48**, 337–361.

Savaş, S., Batta, R., and Nagi, R. (2001). Finite-size facility placement in the presence of barriers to rectilinear travel. Working paper, Dept. of Industrial Engineering, University at Buffalo, NY.

Schandl, B. (1998). On some properties of gauges. Technical Report 662, Department of Mathematical Sciences, Clemson University, SC.

Schöbel, A. (1999). *Locating Lines and Hyperplanes – Theory and Algorithms*. Kluwer Academic Publishers, Dordrecht.

Schwartz, J.T. and Sharir, M. (1990). Algorithmic motion planning in robotics. In *Vol. A of Handbook of Theoretical Computer Science, Ch. 8*. Elsevier.

Schwartz, J.T. and Yap, C.-K. (1987). *Algorithmic and Geometric Aspects of Robotics*. Vol. 1 of Advances in Robotics, Lawrence Erlbaum Associates.

Shamos, M.I. and Hoey, D. (1975). Closest-point problems. *Annals of the Institute of Electrical and Electronics Engineers*, **75**, 151–162.

Shermer, T.C. (1992). Recent results in art galleries. *Proc. IEEE*, **80**, 1384–1399.

Smith, D.R. (1974). *Variational Methods in Optimization.* Prentice-Hall.

Sprau, M. (1999). *Centerprobleme mit Barrieren in der l_2-Metrik – Eine Simulation mit Wasserwellen.* Staatsexamensarbeit am Fachbereich Mathematik der Universität Kaiserslautern.

Steuer, R.E. (1986). *Multiple Criteria Optimization – Theory, Computation and Application.* John Wiley, New York.

Suri, S. (1986). A linear time algorithm for minimum link paths inside a simple polygon. *Computational Vision Graphics Image Processes*, **35**, 99–110.

Tammer, C. and Tammer, K. (1991). Generalization and sharpening of some duality relations for a class of vector optimization problems. *Zeitschrift für Operations Research*, **35**, 249–265.

TenHuisen, M.L. and Wiecek, M.M. (1996). An augmented Lagrangian scalarization for multiple objective programming. In *Proceedings of the MOPGP'96 Conference, Malaga, Spain.*

Thisse, J.-F., Ward, J.E., and Wendell, R.E. (1984). Some properties of location problems with block and round norms. *Operations Research*, **32**, 1309–1327.

Üster, H. and Love, R. (1998a). Application of weighted sums of order p to distance estimation. Technical Report 427, McMaster University, Faculty of Business, Hamilton, Ontario, Canada.

Üster, H. and Love, R. (1998b). On the directional bias of the l_{bp}-norm. Technical Report 428, McMaster University, Faculty of Business, Hamilton, Ontario, Canada.

Vaccaro, H. (1974). *Alternative Techniques for Modelling Travel Distance.* Master's thesis in Civil Engineering, Massachusetts Institute of Technology, USA.

Valentine, F.A. (1964). *Convex Sets.* McGraw-Hill, New York.

Van Der Stappen, A.F. (1994). *Motion Planning Amidst Fat Obstacles.* Ph.D. thesis, University of Utrecht.

Van Laarhoven, P.J.M. and Aarts, E.H.L. (1987). *Simulated Annealing: Theory and Applications.* D. Reidel, Dordrecht.

Verbarg, K. (1996). *Spatial Data Structures and Motion Planning in Sparse Geometric Scenes.* Ph.D. thesis, Universität Würzburg, Germany.

Viegas, J. and Hansen, P. (1985). Finding shortest paths in the plane in the presence of barriers to travel (for any l_p-norm). *European Journal of Operational Research*, **20**, 373–381.

Wang, S.-J., Bhadury, J., and Nagi, R. (2002). Supply facility and input/output point locations in the presence of barriers. *Computers and Operations Research*, **29**, 685–699.

Wangdahl, G., Pollock, S., and Woodward, J. (1974). Minimum trajectory pipe routing. *Journal of Ship Research*, **18**, 46–49.

Wanka, G. (1991). Duality in vectorial control approximation problems with inequality restrictions. *Optimization*, **22**, 755–764.

Wanka, G. (1995). Restricted vectorial location problems. In D. Schweigert, editor, *Methods of Multicriteria Decision Theory, Proc. of the 5th Workshop of the DGOR-Working Group Multicriteria Optimization and Decision Theory*, pages 93–100.

Warburton, A. (1983). Quasiconcave vector maximization: Connectedness of the sets of Pareto-optimal and weak Pareto-optimal alternatives. *Journal of Optimization Theory and Applications*, **40**, 537–557.

Ward, J.E. and Wendell, R.E. (1980). A new norm for measuring distance which yields linear location problems. *Operations Research*, **28**, 836–844.

Ward, J.E. and Wendell, R.E. (1985). Using block norms for location modeling. *Operations Research*, **33**, 1074–1090.

Weißler, A. (1999). *General Bisectors and Their Application in Planar Location Theory*. Shaker Verlag, Aachen.

Weiszfeld, E.V. (1937). Sur le point pour lequel la somme des distances de n points donnés est minimum. *Tohoku Mathematical Journal*, **43**, 335–386.

Welzl, E. (1985). Constructing the visibility graph for n line segments in $O\left(n^2\right)$ time. *Information Processing Letters*, **20**, 167–171.

Wendell, R.E. and Hurter, A.P. (1973). Location theory, dominance and convexity. *Operations Research*, **21**, 314–320.

Wendell, R.E., Hurter, A.P., and Lowe, T.J. (1973). Efficient points in location problems. *AIEE Transactions*, **9**, 238–246.

Wesolowsky, G.O. (1993). The Weber problem: History and perspectives. *Location Science*, **1**, 5–23.

White, D.J. (1982). Dominance and optimal location. *European Journal of Operational Research*, **9**, 309.

Widmayer, P., Wu, Y.F., and Wong, C.K. (1987). On some distance problems in fixed orientations. *SIAM Journal on Computing*, **16**, 728–746.

Witzgall, C. (1964). Optimal location of a central facility: Mathematical models and concepts. Technical Report 8388, National Bureau of Standards, Washington D.C.

Yang, X.Q. and Goh, C.J. (1997). A method for convex curve approximation. *European Journal of Operational Research*, **97**, 205–212.

Index